AS/A-LEVEL

STUDENT GUIDE

OCR

# Geography

## Landscape systems

## Changing spaces; making places

Andy Palmer and Peter Stiff

Hodder Education, an Hachette UK company, Blenheim Court, George Street, Banbury, Oxfordshire OX16 5BH

*Orders*

Bookpoint Ltd, 130 Park Drive, Milton Park, Abingdon, Oxfordshire OX14 4SB

tel: 01235 827827

fax: 01235 400401

e-mail: education@bookpoint.co.uk

Lines are open 9.00 a.m.–5.00 p.m., Monday to Saturday, with a 24-hour message answering service. You can also order through the Hodder Education website: www.hoddereducation.co.uk



ISBN 978-1-4718-6402-5

First printed 2016

Impression number 5 4 3

Year 2018 2017



Cover photo: Richard Carey/Fotolia. Other photos by the authors and Michael Raw

Typeset by Integra Software Services Pvt Ltd, Pondicherry, India

Printed in Dubai

# Contents



## Content Guidance



## Questions & Answers

# Getting the most from this book

**Exam tips**

Advice on key points in the text to help you learn and recall content, avoid pitfalls, and polish your exam technique in order to boost your grade.

**Knowledge check**

Rapid-fire questions throughout the Content Guidance section to check your understanding.

**Knowledge check answers**

1 Turn to the back of the book for the Knowledge check answers.

**Summaries**

- Each core topic is rounded off by a bullet-list summary for quick-check reference of what you need to know.

**Exam-style questions**

**Commentary on the questions**

Tips on what you need to do to gain full marks, indicated by the icon ⓔ

**Sample student answers**

Practise the questions, then look at the student answers that follow.

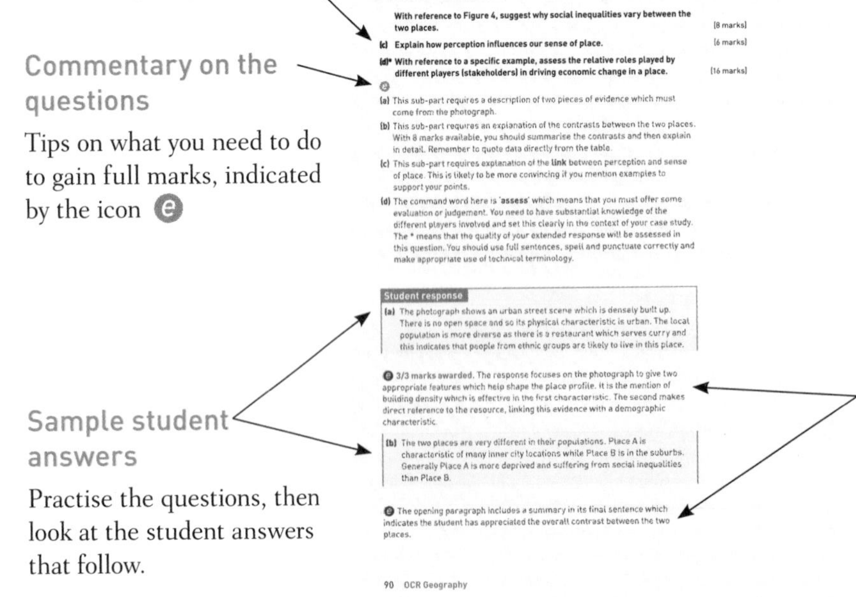

Questions & Answers

With reference to Figure 4, suggest why social inequalities vary between the two places. [8 marks]

(c) Explain how perception influences our sense of place. [6 marks]

(d)* With reference to a specific example, assess the relative roles played by different players (stakeholders) in driving economic change in a place. [16 marks]

ⓔ

(a) This sub-part requires a description of two pieces of evidence which must come from the photograph.

(b) This sub-part requires an explanation of the contrasts between the two places. With 8 marks available, you should summarise the contrasts and then explain in detail. Remember to quote data directly from the table.

(c) This sub-part requires explanation of the **link** between perception and sense of place. This is likely to be more convincing if you mention examples to support your points.

(d) The command word here is '**assess**' which means that you must offer some evaluation or judgement. You need to have substantial knowledge of the different players involved and set this clearly in the context of your case study. The * means that the quality of your extended response will be assessed in this question. You should use full sentences, spell and punctuate correctly and make appropriate use of technical terminology.

**Student response**

(a) The photograph shows an urban street scene which is densely built up. There is no open space and so its physical characteristic is urban. The local population is more diverse as there is a restaurant which serves curry and this indicates that people from ethnic groups are likely to live in this place.

ⓔ 3/3 marks awarded. The response focuses on the photograph to give two appropriate features which help shape the place profile. It is the mention of building density which is effective in the first characteristic. The second makes direct reference to the resource, linking this evidence with a demographic characteristic.

(b) The two places are very different in their populations. Place A is characteristic of many inner city locations while Place B is in the suburbs. Generally Place A is more deprived and suffering from social inequalities than Place B.

ⓔ The opening paragraph includes a summary in its final sentence which indicates the student has appreciated the overall contrast between the two places.

90 OCR Geography

**Commentary on sample student answers**

Read the comments (preceded by the icon ⓔ) showing how many marks each answer would be awarded in the exam and exactly where marks are gained or lost.

# About this book

Much of the knowledge and understanding needed for AS and A-level geography builds on what you have learned for GCSE geography, but with an added focus on geographical skills and techniques, and concepts. This guide offers advice for the effective revision of **Landscape Systems** (Coastal landscapes, Glaciated landscapes and Dryland landscapes), and **Changing Spaces; Making Places**, which all students need to complete.

The external exam papers test your knowledge and application of these aspects of physical and human geography. More information on this is given in the Questions & Answers section at the back of this book. To be successful in these topics you have to understand:

- the key ideas of the content
- the nature of the assessment material — by reviewing and practising sample structured questions
- how to achieve a high level of performance within the examination.

This guide has two sections:

**Content Guidance** — this section summarises some of the key information that you need to know to be able to answer the examination questions with a high degree of accuracy and depth. In particular, the meaning of key terms is made clear and some attention is paid to providing details of case study material to help to meet the spatial context requirement within the specification. Students will also benefit from noting the **Exam tips** that will provide further help in determining how to learn key aspects of the course. **Knowledge check** questions are designed to help learners to check their depth of knowledge – why not get someone else to ask you these?

**Questions & Answers** — this section includes several sample questions similar in style to those you might expect in the exam. There are some sample student responses to these questions as well as detailed analysis, which will give further guidance in relation to what exam markers are looking for to award top marks.

The best way to use this book is to read through the relevant topic area first before practising the questions. Only refer to the answers and examiner comments after you have attempted the questions. **Note:** Throughout the OCR specification the following three terms are used to classify countries. These terms are the ones used by the International Monetary Fund (IMF). The IMF regularly reappraises which group a country is placed in and adjusts its lists accordingly.

| | |
|---|---|
| **Advanced countries (ACs)** | Countries which share a number of important economic development characteristics including well-developed financial markets, high degrees of financial organisation linking demand and supply of capital, goods and information, and diversified economic structures with rapidly growing service sectors. About 30 countries are in this group. |
| **Emerging and developing countries (EDCs)** | Countries which neither share all the economic development characteristics required to be an AC nor are eligible for the Poverty Reduction and Growth Trust, an IMF plan to provide financial support to developing countries. About 80 countries are in this group. |
| **Low-income developing countries (LIDCs)** | Countries which are eligible for the Poverty Reduction and Growth Trust. About 70 countries are in this group. |

# Content Guidance

This section outlines the following areas of the OCR AS and A-level geography specifications:

- Landscape systems: Coastal landscapes
- Landscape systems: Glaciated landscapes
- Landscape systems: Dryland landscapes
- Changing spaces; Making places

# Landscape systems

## Coastal landscapes

### Coastal landscapes as systems

Coastal landscapes can be viewed as systems. A **system** is a set of interrelated elements comprising components (stores) and processes (links) that are connected together to form a working unit or unified whole. Coastal landscape systems store and transfer energy and material.

#### The components of open systems

Coastal landscape systems are open systems. This means that energy and matter can be transferred from neighbouring systems as an input. It can also be transferred to neighbouring systems as an output.

- **Inputs** include kinetic, thermal and potential energy; and material from marine deposition, weathering and mass movement.
- **Outputs** include marine and wind erosion as well as evaporation.
- **Throughputs** consist of stores, such as sediment on a beach, and flows (transfers), such as longshore drift, moving sediment.

**Exam tip**

It is important to be able to distinguish between the components of a system, but also to understand how the components are linked to each other.

**Knowledge check 1**

What are the three different forms of energy in coastal landscape systems?

#### System feedback in coastal landscapes

If a coastal system's inputs and outputs are equal, a state of equilibrium exists. This could happen when the rate at which sediment is being added to a beach equals the rate at which it is being lost from the beach and so the beach will stay the same size. If something happens to upset this equilibrium, the system changes in order to restore the equilibrium. This is known as dynamic equilibrium, as the system produces its own response to the disturbance. This is an example of negative feedback.

**Exam tip**

The dynamic element of system feetback is not the change that occurs to disturb the equilibrium, but the response of the system to restore it. Make that clear in exam answers.

#### Sediment cells

A sediment cell is a stretch of coastline and its associated nearshore area within which the movement of coarse sediment, sand and shingle, is largely self-contained. Sediment cells are usually thought of as closed systems, meaning no sediment can be transferred from one cell to another. In reality, some sediment does get transferred between neighbouring cells. There are also many sub-cells of a smaller scale existing within the major cells.

## Coastal landscape systems are influenced by physical factors

A range of physical factors influence processes which shape the coastal landscape. They vary in their importance and influence spatially (from place to place) and temporally (over time). The factors can also be interrelated as one factor may influence another.

### Winds

Wind is a moving force and can erode, transport and deposit, helping to shape coastal landscapes.

The source of energy for coastal erosion and sediment transport is wave action. Waves are formed by the frictional drag of wind moving across the ocean surface. The higher the wind speed and the longer the **fetch**, the larger the waves and the more energy they possess.

### Waves

- Formation: waves are formed by wind, as explained above. Waves formed in open oceans are called swell waves and generally have a long wavelength with a wave period of up to 20 seconds. In contrast, storm waves are locally generated and typically have a shorter wavelength, greater height and a shorter wave period.
- Development: as waves move forward through the water they can change their form. They can also change their height and/or wavelength.
- Breaking: when waves move into shallow water, they change significantly. Friction between the sea floor and the water causes waves to slow down. The height increases, wavelength decreases and successive waves start to bunch up. When water depth is less than $1.3 \times$ wave height, the wave breaks.

There are three types of breaking wave:

- spilling — steep waves breaking onto gently sloping beaches; water spills gently forward as the wave breaks
- plunging — moderately steep waves breaking onto steep beaches; water plunges vertically downwards
- surging — low angle waves breaking onto steep beaches; the wave slides forward.

### Tides

Tidal cycles are the periodic rise and fall of the sea surface. Tides are largely produced by the gravitational pull of the moon.

- The moon pulls the water towards it, creating a high tide, and there is a compensatory bulge on the opposite side of the Earth.
- At locations between the two bulges, there will be a low tide.
- As the moon orbits the Earth, the high tides follow it. The highest tides will occur when the moon, sun and Earth are all aligned and so the gravitational pull is at its strongest. This happens twice each lunar month and results in spring tides with a high tidal range.

**Exam tip**

There are many sub-cells of a smaller scale existing within the major cells. If you use a case study from the UK later in this topic, it is worth knowing which cell, or sub-cells, it includes.

**Fetch** is the distance of open sea over which the wind blows.

**Exam tip**

The speed, direction and frequency of winds all influence wave formation and aeolian processes. Make sure you can explain how.

**Exam tip**

The water in a wave moves in a circular motion, rather than forward, until it breaks, when there is significant forward movement. This is helpful when you are explaining how waves transfer energy to the coast.

**Knowledge check 2**

What causes waves to break?

- Also twice a month, the moon and the sun are at right-angles to each other and the gravitational pull is therefore at its weakest, producing neap tides with a low range.

**Tidal range** can be a significant factor in the development of coastal landscapes. In enclosed seas tidal ranges are low and so wave action is restricted to a narrow area of land. In places where the coast is funnelled, such as estuaries, tidal range is much higher.

**Tidal range** is the difference between the water level at high tide and at low tide.

## Geology

**Lithology** refers to the physical and chemical composition of rocks. Some rock types have a weak lithology, with little resistance to erosion, weathering and mass movements. This is because the bonds between the particles that make up the rock are quite weak, as in clay. Others, such as basalt, made of dense interlocking crystals, are very resistant. Some, such as chalk, are largely composed of calcium carbonate, and so soluble in weak acids, making them prone to chemical weathering by carbonation.

**Structure** concerns the properties of individual rock types such as jointing, bedding and faulting which affects the permeability of rocks. In porous rocks, such as chalk, tiny air spaces separate the mineral particles. These pores can absorb and store water, known as primary permeability. Carboniferous limestone is also permeable, because of its many joints. This is known as secondary permeability.

**Exam tip**

Make sure you can explain how tidal range influences where wave action occurs, the weathering processes that happen on land exposed between tides and the potential scouring effect of waves along coasts with a high tidal range.

## Ocean currents

Ocean currents are generated by the Earth's rotation and by convection, and are set in motion by the movement of winds across the water surface. Warm ocean currents transfer heat-energy from low latitudes towards the poles. Cold ocean currents do the opposite, moving cold water from polar regions towards the Equator.

The strength of the current itself has a limited impact on coastal landscape processes, but the transfer of heat-energy can be significant, as it directly affects air temperature and, therefore, the subaerial processes of weathering and mass movement.

**Exam tip**

When explaining the influence of geology on coastal landscapes, it can be useful to refer to their planform. Rock outcrops that are uniform, or run parallel to the coast, tend to produce straight coastlines, known as concordant coasts. Where rocks lie at right-angles to the coast they create discordant coastlines, which are much more irregular.

# Sources of coastal sediment

Coastal sediment comes from a variety of sources.

## Terrestrial

Terrestrial sources of coastal sediment contribute to the following processes.

- Fluvial deposition: the origin of terrestrial sediment is the erosion of inland areas by water, wind and ice as well as subaerial processes of weathering and mass movement. This sediment is then transported to the coast by rivers, which deposit it at their mouths as they lose energy.
- Marine erosion: cliff erosion by waves is a significant source of coastal sediment. The erosion of weak cliffs in high-energy wave environments can contribute as much as 70% of the overall material supplied to beaches, although usually it is much less. Some of this sediment may be large rocks and boulders, often from mass movement on undercut cliffs.
- **Aeolian** deposition: winds carry fine particles and deposit them as they lose energy,
- Longshore drift: this can supply sediment to one coastal area by moving it along the coast from adjacent areas.

**Knowledge check 3**

Why is chalk classified as a porous rock?

**Aeolian** means wind related.

## Offshore

Offshore sources contribute sediment in the following ways.

- Waves bring sediment onshore from offshore locations. Marine deposition is when waves deposit sediment as they lose energy, typically after they have broken.
- Tides and currents do the same.
- Wind also blows sediment from offshore locations, including exposed sand bars.

**Exam tip**

Exam questions may specifically ask about **either** physical **or** human sources. Make sure you can distinguish between the two.

## Human

Beach nourishment involves human management of beaches. Sand is often added to beaches as a feature of coastal management and protection against erosion. Sand brought in from external sources may be pumped onshore or brought by lorry.

# How are coastal landforms developed?

Coastal landforms develop due to a variety of interconnected climatic and geomorphic processes.

## The influence of flows of energy and materials on geomorphic processes

There are a number of geomorphic processes that occur in coastal landscapes:

### *Weathering*

**Weathering** happens everywhere, but different types of weathering are more or less significant in different types of landscape.

**Weathering** is the breakdown and decay of rock through exposure to the Earth's atmosphere, organisms and water.

#### *Physical (mechanical) weathering*

This involves a number of different processes.

- Freeze–thaw: water enters cracks/joints and expands by nearly 10% when it freezes. This exerts pressure on the rock causing it to split or causing pieces to break off.
- Pressure release: when overlying rocks are removed by weathering and erosion, the underlying rock expands and fractures parallel to the surface.
- Thermal expansion: rocks expand when heated and contract when cooled. If they are subjected to frequent cycles of temperature change then the outer layers may crack and flake off.
- Salt crystallisation: solutions of salt in sea water can enter pore spaces in porous rocks. The salts precipitate, forming crystals, and the growth of these crystals creates stress in the rock causing it to disintegrate.

#### *Chemical weathering*

Chemical weathering can be of several different types.

- Oxidation: some minerals, especially iron, in rocks react with oxygen, either in the air or in water. The rock becomes soluble under strongly acidic conditions and the original structure is destroyed.
- Carbonation: rainwater is a weak carbonic acid. This reacts with calcium carbonate in rocks such as limestone to produce calcium bicarbonate, which is soluble.
- Solution: some minerals are soluble in water and as they dissolve they weaken the structure of a rock.

- Hydrolysis: this is a chemical reaction between rock minerals and water. Silicates combine with water producing secondary minerals such as clays.
- Hydration: water molecules added to rock minerals create new minerals of a larger volume. This happens to anhydrite, forming gypsum.

### *Biological weathering*

Biological weathering involves the actions of organisms such as plants and animals.

- Tree roots grow into cracks or joints in rocks and exert outward pressure, causing rock to split. Burrowing animals may have a similar effect.
- Organic acids may be secreted by molluscs, released during the decay of organic matter or released by algae. These acids can react with rock minerals.

**Exam tip**

When answering exam questions about the effects of weathering, show that you understand the changes that happen to rocks. Physical weathering tends to produce smaller fragments of the same material, whereas chemical weathering produces chemically altered substances.

## *Mass movement*

**Mass movement** processes may move material on slopes in coastal landscapes, especially on cliff faces.

- Rock fall: on cliffs of 40° or more rocks may become detached by physical weathering processes. These rocks then fall to the foot of the cliff under gravity.
- Slides: these may be linear, with movement along a straight-line slip plane, such as a fault or a bedding plane between layers of rock, or rotational, with movement taking place along a curved slip plane. Rotational slides are also known as slumps.

**Mass movement** is the downslope movement of material under gravity, without the aid of a moving force.

## *Wave (marine) processes*

### *Erosion*

Most erosion in coastal landscapes is carried out by breaking waves.

- Abrasion (or corrasion) is when waves carrying rock particles scour the coastline, rock rubbing against rock.
- Attrition occurs when rock particles, carried in waves, collide with each other and with coastal rocks, becoming worn away.
- Hydraulic action occurs when waves break against a cliff face, and air and water trapped in cracks and crevices becomes compressed. As the wave recedes the pressure is released, the air and water suddenly expand and the crack is widened.
- Pounding occurs when the mass of a breaking wave exerts pressure on the rock causing it to weaken.
- Solution (or corrosion) involves minerals such as magnesium carbonate minerals in coastal rock dissolving in sea water.

**Knowledge check 4**

What is the difference between mass movement and transportation?

### *Transportation*

Sediment can be moved within the sea or moved towards to coast, mainly by wave action.

- Solution: minerals have been dissolved into the mass of moving water.
- Suspension: small particles of sand, silt and clay can be carried by currents. This accounts for the brown or muddy appearance of some sea water.
- Saltation: a series of irregular movements of material, which is too heavy to be carried continuously in suspension. Turbulent flow may pick up particles and carry them for a short distance only to drop them again.
- Traction: the largest particles in the load may be pushed along the sea floor by the force of the flow. This is also called rolling, although the movement is seldom continuous.

Sediment can also be moved along the coast by longshore drift. This occurs when waves approach the coast at an angle due to the direction of the dominant wind. After breaking, the swash carries particles diagonally up the beach. The backwash moves them perpendicularly back down the beach due to gravity. If this movement is repeated, the net result is a movement of material along the beach.

### Deposition

Material is deposited when there is a loss of energy caused by a decrease in velocity and/or volume of water.

Deposition tends to take place in coastal landscapes:

- where the rate of sediment accumulation exceeds the rate of removal
- when waves slow down immediately after breaking
- at the top of the swash, where for a brief moment the water is no longer moving
- during the backwash, when water percolates into the beach material
- in low-energy environments, those sheltered from winds and waves, such as estuaries.

**Exam tip**

When answering exam questions about the effects of wave processes, you should show that you understand the roles of energy and material.

## Fluvial (river) processes

The processes associated with rivers, especially at their mouths, contribute to the development of coastal landforms.

### Erosion

Fluvial erosion upstream is the main source of a river's sediment load. Rivers use similar erosional processes to waves. Sediment is also derived from weathering and mass movement processes on valley sides which then enters the channel.

### Transportation

Rivers transport sediment in the same way as waves- by traction, suspension, saltation and solution.

### Deposition

As rivers enter the sea, there is a noticeable reduction in their velocity and so energy is reduced. Some, or all, of the river's sediment load is deposited. As the reduction in energy is progressive, deposition is sequential, with the largest particles being deposited first and the finest being carried further out to sea. Deposition also results from **flocculation**.

**Flocculation** is the process by which salt causes the aggregation of minute clay particles into larger masses that are too heavy to remain suspended in water.

## Aeolian (wind) processes

Winds blowing onshore can be effective in the development of coastal landforms.

### Erosion

Wind is able to pick up sand particles and move them. This enables abrasion to occur, as it does in waves. Particles being carried are also subjected to attrition.

### Transportation

With the exception of solution, moving air is able to transport material using the same mechanisms as water moving in rivers and waves.

**Knowledge check 5**

What is the difference between erosion and weathering?

### Deposition

Material carried by wind will be deposited when the wind speed falls, usually due to surface friction, resulting in a loss of energy.

> **Exam tip**
>
> Fluvial and aeolian processes may play a less important role in the development of coastal landforms than wave processes, but they do contribute. You may be asked to compare their roles.

## The formation of distinctive landforms

Distinctive landforms are formed by erosion and deposition.

### Erosional landforms

#### Bays and headlands

Bays and headlands typically form adjacent to each other, usually due to the presence of bands of rock of differing resistance to erosion. If these rock outcrops lie perpendicular to the coastline, the weaker rocks are eroded more rapidly to form bays while the more resistant rocks remain between bays as headlands. This results in the formation of a **discordant** coastline.

Rocks lying parallel to the coastline produce a **concordant** coastline. If the most resistant rock lies on the seaward side it protects any weaker rocks inland from erosion. The resultant coastline is quite straight and even. However, even in this situation small bays or coves may occasionally be eroded at points of weakness, such as fault lines.

#### Cliffs and shore platforms

When waves break repeatedly on coastlines, undercutting can occur between the high and low tide levels where it forms a **wave-cut notch**. Continued undercutting weakens support for the rock above which eventually collapses producing a steep profile and a cliff. The regular removal of debris at the foot of the cliff by wave action ensures that the cliff profile remains relatively steep and that the cliffs retreat inland parallel to the coast.

> **Exam tip**
>
> Bays and headlands, and cliffs and shore platforms, tend to be found together, as pairs, and so it is difficult to explain the development of one without explaining the other as well.

As the sequence of undercutting, collapse and retreat continues, the cliff becomes higher. At its base, a gently sloping shore platform is cut into the solid rock. Although superficially appearing to be flat and even, **shore platforms** are often deeply dissected by abrasion due to the large amount of rock debris that is dragged across the surface by wave action. They are also affected by weathering processes at low tide. Mass movement from the cliff may also leave large rocks on the platform.

#### Geos and blow holes

**Geos** are narrow, steep-sided inlets. Weak points are eroded more rapidly, by wave action, than the more resistant rock around them. Hydraulic action may be particularly important in forcing air and water into joints and weakening the rock.

If part of the roof of a cave collapses, it may form a vertical shaft that reaches the cliff top. This is a **blow hole**. In storm conditions large waves may force spray out of the blow hole as plumes of white, aerated water. Both geos and blow holes may form from collapsed mineshafts.

#### Caves, arches, stacks and stumps

Due to **wave refraction**, energy is concentrated on the sides of headlands.

> **Wave refraction** is the re-orientation of a wave so that it approaches a shoreline at a more perpendicular angle.

Any points of weakness, such as faults or joints, are exploited by erosion processes and a small cave may develop on one, or even both, sides of the headland. Wave erosion is concentrated between high and low tide levels and it is here that **caves** form. If a cave enlarges to such an extent that it extends through to the other side of the headland, possibly meeting another cave, an **arch** is formed. Continued erosion widens the arch and weakens its support. Aided by weathering processes, the arch may collapse leaving an isolated **stack** separated from the headland. Further erosion at the base of the stack may eventually cause further collapse leaving a small, flat portion of the original stack as a **stump**. This may only be visible at low tide.

**Knowledge check 6**

Why can waves erode both sides of a headland?

## Depositional landforms

### Beaches

Beaches are the most common landform of deposition and result from the accumulation of material deposited between the lowest tides and the highest storm waves.

### Types of beach

Sand produces beaches with a gentle gradient because its small particle size means that it becomes compact when wet, allowing little percolation during backwash. As little energy is lost to friction, and little volume is lost to percolation, material is carried back down the beach rather than being left at the top, resulting in a gentle gradient.

**Shingle**, a mix of pebbles and small-to-medium sized cobbles, produces steeper beaches because swash is stronger than backwash so there is a net movement of shingle onshore. Shingle may make up the upper part of the beach where rapid percolation due to larger air spaces means that little backwash occurs and so material is left at the top of the beach.

Storm waves hurl pebbles and cobbles to the back of the beach forming a **storm beach** or storm ridge.

### Beach features

**Berms** are small ridges that develop at the position of the mean high-tide mark, resulting from deposition at the top of the swash.

**Cusps** are small, semi-circular depressions; temporary features formed by a collection of waves reaching the same point and when swash and backwash have similar strength.

Further down the beach, **ripples** may develop in the sand due to the orbital movement of water in waves.

### Spits

**Spits** are long, narrow beaches of sand or shingle that are attached to the land at one end and extend across a bay, estuary or indentation in a coastline. They are generally formed by longshore drift occurring in one dominant direction which carries beach material to the end of the beach and then beyond into the open water. The end of the spit often becomes recurved as a result of wave refraction around the end of the spit and/or a secondary wind/wave direction. Over time spits may continue to grow and a number of recurves may develop.

In the sheltered area behind the spit, deposition will occur as wave energy is reduced. The silt and mud deposited build up and eventually salt tolerant vegetation may colonise leading to the formation of a **salt marsh**.

### Onshore bars

**Onshore bars** can develop if a spit continues to grow across an indentation, such as a cove or bay, in the coastline until it joins onto the land at the other end. This forms a lagoon of **brackish** water on the landward side.

**Brackish** water is a mixture of salt and fresh water.

Some may also have been formed by the onshore movement of sediment during the post-glacial sea-level rise that ended about 6,000 years ago.

### Tombolos

**Tombolos** are beaches that connect the mainland to an offshore island. They are often formed from spits that have continued to grow seawards until they reach and join an island. However, as with onshore bars, they may also have been formed by the onshore movement of sediment during the post-glacial sea-level rise.

### Salt marshes

Salt marshes are features of low-energy environments, such as estuaries and on the landward side of spits, and are vegetated areas of deposited silts and clays. Deposition of fine sediment occurs as rivers lose energy when they slow upon entering the sea and also due to flocculation. They are subjected to twice-daily inundation and exposure as tides rise and fall. Salt-tolerant plant species such as eel grass and spartina help trap sediment, gradually helping to increase the height of the marsh.

**Knowledge check 7**

What is flocculation?

Extensive networks of small, steep-sided channels, or creeks, drain the marsh at low tide and provide routes for water to enter the salt marsh as the tide rises. Between the creeks, shallow depressions are often found. These trap water when the tide falls, and these areas of saltwater, called **saltpans**, are often devoid of any vegetation.

### Deltas

**Deltas** are large areas of sediment found at the mouths of rivers. They form when rivers and tidal currents deposit sediment at a faster rate than waves and tides can remove it. Deltas are criss-crossed by a branching network of distributaries. Overloaded with sediment, deposition in the channel forms bars which causes the channel to split into two. This produces two channels with reduced energy levels, and so more deposition and further dividing occurs. These channels may be lined by **levées** on their banks.

**Levées** are raised embankments formed by deposition during flooding.

The structure of deltas usually consists of three distinctive components.

a The upper delta plain: furthest inland, beyond the reach of tides and composed entirely of river deposits.

b The lower delta plain: in the inter-tidal zone, regularly submerged and composed of both river and marine deposits.

c The submerged delta plain: below mean low water mark, composed mainly of marine sediments and represents the seaward growth of the delta.

**Exam tip**

Each of the landforms of deposition listed should be known and their development understood. Questions may ask for explanation of their formation, but examples will not be required. However, if examples are used, and aid the answer, they can be credited.

# Coastal landforms are interrelated

Coastal landforms are interrelated and together make up characteristic landscapes.

## Case studies

For this part of the specification, you are required to have two case studies. At least one of these case studies should be from beyond the UK.

The case studies must include one high-energy coastline and one low-energy coastline.

### High-energy coastline

This case study could be a rocky coastline, such as the Flamborough Head area of Yorkshire, Land's End in Cornwall or Port Campbell, south-eastern Australia.

For your case study, you should be able to illustrate the following.

- The physical factors which influence the formation of landforms within the landscape system. These are likely to include geology, winds, tides, waves, sediment cells and sediment sources.
- The interrelationship of a range of landforms within the landscape system. These might include cliffs, shore platforms, bays (and their beaches), geos, blow holes, caves, arches, stacks and stumps.
- How and why the landscape system changes over time. This should include short-term changes such as cliff or arch collapse, medium-term changes such as seasonal changes in beach profiles, and long-term changes such as shore platform development.

You should be able to explain:

- how the physical factors have influenced the geomorphic processes
- how the geomorphic processes have led to the development of the landforms
- how the landforms interrelate with each other
- how and why the landscape system changes over time
- how and why energy and material are transferred through the coastal landscape system.

To do this successfully, you should know:

- names, locations and approximate scale of a range of landforms
- the rock types forming the geology of the area, and their relative resistance to geomorphic processes
- the prevailing and dominant wind directions (which may be the same), the length of the fetch and the tidal range.

### Low-energy coastline

This case study could be an estuarine or deltaic coastline, such as the Nile Delta in Egypt, the Humber Estuary in Yorkshire or the mangrove swamps of Vietnam.

For your case study, you should be able to illustrate the following.

- The physical factors which influence the formation of landforms within the landscape system. These are likely to include geology, winds, tides, waves, sediment cells and sediment sources.
- The interrelationship of a range of landforms within the landscape system. These might include beaches, spits, onshore bars, tombolos and salt marshes.
- How and why the landscape system changes over time. This should include short-term changes such as the breaching of onshore bars, medium-term changes such as seasonal changes in beach profiles, and long-term changes such as spit or tombolo development.

You should be able to explain:

- how the physical factors have influenced the geomorphic processes
- how the geomorphic processes have led to the development of the landforms
- how the landforms interrelate with each other
- how and why the landscape system changes over time
- how and why energy and material are transferred through the coastal landscape system.

To do this successfully, you should know:

- names, locations and approximate scale of a range of landforms
- the rock types forming the geology of the area, and their relative resistance to geomorphic processes
- the prevailing and dominant wind directions (which may be the same), the length of the fetch and the tidal range.

# Coastal landforms evolve as climate changes

How do coastal landforms evolve over time as climate changes?

## Emergent coastal landscapes

Emergent coastal landscapes form as sea level falls. Landforms in emergent landscapes are influenced by falling sea level due to a cooling climate.

### *Climate change and the resultant sea-level fall*

About 130,000 years ago, during the Tyrrhenian inter-glacial period, global mean annual temperatures were almost 3°C higher than today and sea level was about 20 m above today's position. Temperatures then fell during the onset of the Riss glacial period, reaching a minimum about 7°C lower than today about 108,000 years ago. As a result of this temperature decrease, less water was returned to the ocean store and sea levels dropped by over 100 m, making them about 83 m lower than the present day.

This happened because a decrease in global temperature leads to more precipitation being in the form of snow. Eventually this snow turns to ice and so water is stored on the land in solid form rather than being returned to the ocean store as liquid. The result is a reduction in the volume of water in the ocean store and a worldwide fall in sea level. Also, as temperatures fall, water molecules contract leading to an increased density and a reduced volume.

### *Emergent landforms*

Landforms shaped by wave processes during times of high sea level are left exposed when sea level falls. As a result they may be found well inland, some distance from the modern coastline.

#### *Raised beaches and abandoned cliffs*

**Raised beaches** are areas of former shore platforms that are found at a higher level than the present sea level. They are often found a distance inland from the present coastline. Behind the beach, along emergent coastlines, it is not uncommon to find **abandoned cliffs** with wave-cut notches, caves and even arches and stacks.

**Exam tip**

Particularly for essay writing, it would be helpful to be able to draw a sketch map of the area with the key features labelled.

During inter-glacial periods when sea levels were much higher than today's, erosion rates were higher. This was due to the deeper water allowing wave energy to be greater, with less being lost to sea-floor friction. Wave erosion also occurred at higher elevations, forming active cliffs and shore platforms. When sea levels fell during the next glacial period, these landforms were left behind, further inland of the new, lower sea-level position.

#### *Marine terraces*

Sometimes sea level falls in a series of stages. Should that happen, then marine erosion is able to develop a number of **marine terraces**, appearing as steps in the landscape; each one represents a period of stable temperatures and sea level during which wave action had sufficient time to act significantly on the landscape to produce a flat, eroded terrace. Although these landforms are associated with a decrease in sea level due to a cooling climate, they can also be related to rising land levels.

### *Landform modification*

After their emergence, these landforms are no longer affected by wave processes. However, they continue to be affected by weathering and mass movement.

In the post-glacial period, warmer and wetter conditions have led to the development of vegetation cover on many such landforms, often making them more difficult to recognise. With further warming of the climate predicted for the future, continued degradation is likely to occur with higher rates of chemical weathering and mass movement. Chemical weathering on the raised beach may also become more significant in the warmer, wetter climatic conditions.

If temperatures increase sufficiently, the associated sea-level rise could lead to these emergent landforms again being found much closer to, or even at, the coastline. They would then be subjected to wave processes once more.

## Submergent coastal landscapes

Submergent coastal landscapes form as sea level rises. Landforms in submergent landscapes are influenced by rising sea level due to a warming climate.

### *Climate change and the resultant sea-level rise*

At the end of the Würm glacial period, which happened about 25,000 years ago, temperatures were about 9°C lower than today and sea level was about 90 m lower than the present. Since then temperatures and sea level have risen to their present level. This period of significant sea-level rise is known as the Flandrian Transgression.

This happened because an increase in global temperature leads to higher rates of melting of the ice stored on the land in glaciers. As a consequence there is a global increase in the volume of water in the ocean store and a consequent rise in sea level. Also, as temperatures rise, water molecules expand and this also leads to an increased volume of sea water.

**Exam tip**

With both falling and rising sea level, the specification requires the study of **a** specific previous time period, so you should be able to provide data, as evidence of the temperature and sea-level changes.

## Submergent landforms

Landforms shaped by wave processes during times of lower sea level are left submerged, or drowned, when sea level rises.

### Rias

**Rias** are submerged river valleys, formed as sea level rises. The lowest part of the river's course and the floodplains alongside the river may be completely drowned, but the higher land forming the tops of the valley sides and the middle and upper part of the river's course remain exposed. In cross-section rias have relatively shallow water becoming increasingly deep towards the centre. The exposed valley sides are quite gently sloping. In long-section they exhibit a smooth profile and water of uniform depth. In plan view they tend to be winding, reflecting the original route of the river and its valley, formed by fluvial erosion within the channel and subaerial processes on the valley sides.

### Fjords

**Fjords** are submerged glacial valleys. They have steep, almost cliff-like, valley sides and the water is uniformly deep, often reaching over 1,000 m. The U-shaped cross-section reflects the original shape of the glacial valley itself. They consist of a glacial rock basin with a shallower section at the end known as the **threshold**. This results from lower rates of erosion at the seaward end of the valley where the ice thinned in warmer conditions. They also tend to have much straighter planforms than rias as the glacier would have truncated any interlocking spurs present.

Due to the depth of water that occupied fjords during the Flandrian Transgression, marine erosion rates remained high and in some cases the fjords were further deepened.

**Exam tip**

When answering questions about landform modification, remember that you are speculating about what could happen in the future. You should use appropriate, tentative language, such as 'may' and 'likely to'.

### Shingle beaches

When sea level falls, large areas of 'new' land emerges from the sea. Sediment accumulates on this surface, deposited by rivers and low-energy waves. As sea level rose at the end of the last glacial period, wave action pushed these sediments onshore. In some places they were deposited at the base of former cliff lines, elsewhere they form tombolos and bars.

## Landform modification

Both rias and fjords may be modified by the wave processes acting on their sides at the present-day sea level. The valley sides may also be affected by the operation of subaerial processes in today's climatic conditions or in any different climatic conditions of the future. This may eventually lead to a reduction in the steepness of the valley sides of fjords.

**Knowledge check 8**

Explain two reasons why temperature change causes sea-level change.

With sea levels predicted by some people to rise by a further 0.6 m in the next 100 years, water depth in rias and fjords will increase. Marine erosion is also likely to increase due to stormier conditions and larger waves.

Shingle beaches, being composed of unconsolidated material rather than solid rock, are especially vulnerable to modification.

# Human activity changes coastal landscapes

Human activity intentionally causes change within coastal landscape systems.

## Case study

For this part of the specification, you are required to have a case study of one coastal landscape that is being managed. This could be the same coastal landscape as you used for a case study of high-energy or low-energy coastlines, or a completely different one.

You might have studied a location such as Studland or Sandbanks in Dorset, Holderness in Yorkshire or Santa Barbara in California, USA.

For your case study, you should be able to describe and explain the following.

- The management strategy being implemented and the reason for its implementation. Strategies might include groyne or sea-wall construction, or offshore dredging.
- The intentional impacts on the processes and flows of material and/or energy through the coastal system. This might particularly relate to the sediment budget.
- The effect of these impacts in changing coastal landforms. This might include changes to a beach profile or planform.
- The consequence of these changes on the landscape. This might be through changing a number of landforms, or broad, more generic changes such as the extension of the coastal landscape seawards.

To do this successfully, you should know:

- the name, location and approximate scale of the management strategy
- the date of its implementation
- the processes and flows affected by the strategy, with supporting data
- the changes that occurred to coastal landforms and landscape and the timescale over which the changes occurred.

# Economic development changes coastal landscapes

Economic development unintentionally causes change within coastal landscape systems.

## Case study

For this part of the specification, you are required to have a case study of one coastal landscape that is being used by people. This could be the same coastal landscape as you used for a case study of high-energy or low-energy coastlines, or a completely different one.

For your case study, you should be able to describe and explain the following.

- The economic activity taking place and the reasons for it taking place. This might include the use of the coast for trade routes, port development or tourist developments.
- The unintentional impacts on processes and flows of material and/or energy through the coastal system, such as disturbance to the sediment cell balance.

- The effect of these impacts in changing coastal landforms. This might include changes to rates of erosion and/or deposition.
- The consequence of these changes on the landscape. This might be through changing a number of landforms, or broad, more generic changes such as the retreat of the coastal landscape or the resulting need for coastal management.

To do this successfully, you should know:

- the type, location and approximate scale of the economic development
- the date(s) of its development
- the processes and flows affected by the development, with supporting data
- the changes that occurred to coastal landforms and landscape and the timescale over which the changes occurred.

**Exam tip**

Particularly for essay writing, it would be helpful to be able to draw a sketch map of the area with the key features of the human activity and the affected landforms labelled.

## Summary

- Coastal landscape systems consist of flows of energy and material, and sediment cells.
- Coastal landscape systems are influenced by a range of physical factors: winds, waves, tides, geology and ocean currents.
- There are a variety of sources of sediment supply: terrestrial, offshore and human.
- Flows of energy and materials influence geomorphic processes.
- Distinctive landforms are formed by erosion and by deposition.
- Case studies are required of one high-energy and one low-energy coastline to illustrate: the physical factors which influence the formation of landforms, the interrelationship of a range of landforms and how and why the landscape system changes over time.
- Climate changes have occurred during a previous time period and resulted in sea-level fall and sea-level rise.
- Sea-level fall, sea-level rise and geomorphic processes influence the shaping of landforms. These landforms are influenced by processes associated with present and future climate and sea-level changes.
- Case studies are required of one coastal landscape that is being managed and one coastal landscape that is being used by people.

# Glaciated landscapes

## Glaciated landscapes as systems

Glaciated landscapes can be viewed as systems. A **system** is a set of interrelated elements comprising components (stores) and processes (links) that are connected together to form a working unit or unified whole. Glaciated landscape systems store and transfer energy and material.

### The components of open systems

Glaciated landscape systems are open systems. This means that energy and matter can be transferred from neighbouring systems as an input. It can also be transferred to neighbouring systems as an output.

- **Inputs** include kinetic, thermal and potential energy; and material from glacial deposition, weathering and mass movement.
- **Outputs** include melting as well as evaporation and **sublimation**.
- **Throughputs** consist of stores, such as sediment held in a **glacier**, and flows (transfers), such as basal sliding of ice.

### System feedback in glaciated landscapes

If a glaciated system's inputs and outputs are equal, a state of equilibrium exists. This could happen when the rate at which ice is being added to a glacier equals the rate at which it is being lost from the glacier, and so the glacier will stay the same size. If something happens to upset this equilibrium, the system changes in order to restore the equilibrium. This is known as dynamic equilibrium, as the system produces its own response to the disturbance. This is an example of negative feedback.

### Glacier mass balance

The glacier mass balance, or budget, is the difference between the amount of snow and ice accumulation and the amount of **ablation** occurring in a glacier over a one-year time period.

The upper reaches of the glacier, where accumulation exceeds ablation, is called the **accumulation zone**. Most of the outputs occur at lower levels where ablation exceeds accumulation, in the **ablation zone**. The two zones are notionally divided by the **firn** or equilibrium line where there is a balance between accumulation and ablation.

The annual mass balance can be calculated by subtracting the total ablation for the year from the total accumulation. A positive figure indicates a net gain of ice, increasing the volume of ice and allowing the glacier to advance or grow. A negative figure indicates a net loss of ice through the year. In this situation ablation exceeds accumulation and the glacier will contract and retreat up-valley.

If the amount of accumulation equals the amount of ablation the glacier is in equilibrium and therefore remains stable in its position.

There will often be seasonal variations in the budget with accumulation exceeding ablation in the winter and vice versa in the summer. It is, therefore, possible that there will be some advance during the year even if the net budget is negative and some retreat even when it is positive.

**Sublimation** is the direct change of state from solid (ice) to gaseous (water vapour).

**Exam tip**

It is important to be able to distinguish between the components of a system, but also to understand how the components are linked to each other.

**Knowledge check 9**

What are the three different forms of energy in glaciated landscape systems?

**Exam tip**

The dynamic element of system feedback is not the change that occurs to disturb the equilibrium, but the response of the system to restore it. Make that clear in exam answers.

**Ablation** includes all losses of ice from a glacier, including melting, evaporation, sublimation and iceberg calving.

**Knowledge check 10**

How is the annual mass balance of a glacier calculated?

**Exam tip**

An idea that many fail to appreciate is that even when in retreat, the ice in a glacier may move forwards across the firn line under gravity, so it can appear that a retreating glacier is actually advancing.

# Glaciated landscape systems are influenced by physical factors

A range of physical factors influence processes which shape the glaciated landscape. They vary in their importance and influence spatially (from place to place) and temporally (over time). These factors can also be interrelated as one factor may influence another.

## Climate

Wind is a moving force and as such is able to carry out erosion, transportation and deposition. These **aeolian** processes contribute to the shaping of glaciated landscapes, particularly acting upon fine material previously deposited by ice or meltwater.

**Aeolian** means wind related.

Precipitation totals and patterns are key factors in determining the mass balance of a glacier, as precipitation provides the main inputs of snow, sleet and rain.

Temperature is also a significant factor. If temperatures rise above 0°C, accumulated snow and ice will start to melt and become an output of the system. High altitude glaciers may experience significant periods in the summer months of above zero temperatures and melting, whereas, in high latitude locations, temperatures may never rise above zero and so no melting occurs. This explains why ice sheets are so thick in polar regions, despite very low precipitation inputs.

**Exam tip**

Make sure you can comment on the seasonal variations in the climatic factors.

## Geology

**Lithology** refers to the physical and chemical composition of rocks. Some rock types have a weak lithology, with little resistance to erosion, weathering and mass movement. This is because the bonds between the particles that make up the rock are quite weak, as in clay. Others, such as basalt, made of dense interlocking crystals, are very resistant. Some, such as chalk, are largely composed of calcium carbonate, and so are soluble in weak acids making them prone to chemical weathering by carbonation.

**Structure** concerns the properties of individual rock types such as jointing, bedding and faulting which affects the permeability of rocks. In porous rocks, such as chalk, tiny air spaces separate the mineral particles. These pores can absorb and store water, known as primary permeability. Carboniferous limestone is also permeable, because of its many joints. This is known as secondary permeability.

**Exam tip**

When explaining the influence of geology on glaciated landscapes, it can be useful to refer to how the shape of the landform, such as the cross-section of a trough, is affected by geology.

## Latitude and altitude

Locations at high latitudes, most noticeably beyond the Arctic and Antarctic Circles at 66.5°N and S, tend to have cold dry climates with little seasonal variation. Glaciated landscapes at such latitudes tend to develop under the influence of large, relatively stable ice sheets, such as those of Greenland and Antarctica. These landscapes are quite different to those that develop under

**Knowledge check 11**

Why is chalk classified as a porous rock?

the influence of dynamic valley glaciers in lower latitude, but higher altitude locations, such as the Rocky Mountains and the Himalayas. These locations tend to have higher precipitation inputs, but more variable temperatures and hence more summer melting.

The decrease in temperature with altitude of approximately 0.6°C/100 m means that glaciers are even found near the Equator in the Andes. High altitude locations may also receive more relief precipitation.

### Relief and aspect

Although latitude and altitude are the major controls on climate, relief and **aspect** have an impact on microclimate and the movement of glaciers.

**Aspect** is the direction a slope faces.

The steeper the relief of the landscape, the greater the resultant force of gravity and the more energy a glacier will have to move downslope.

Where air temperature is close to zero and the melting of snow and ice, aspect can have a significant influence on the behaviour of glacier systems. If the aspect of a slope faces away from the general direction of the sun, temperatures are likely to remain below zero for longer, as less solar energy is received, and so less melting occurs. The mass balance of glaciers in such locations will, therefore, tend to be positive, causing them to advance. The reverse is likely to be true in areas with an aspect facing towards to the sun. These differences not only affect the mass balance, but will, as a result, influence the shaping of the landscape. Glaciers with a positive mass balance are more likely to be larger, with greater erosive power and much more erosive than small ones and those in retreat due to a negative mass balance.

## Different types of glacier and glacier movement

### The formation of glacier ice

Glaciers form when temperatures are low enough for snow that falls in one year to remain frozen throughout the year. This means that the following year, fresh snow falls on top of the previous year's snow. Fresh snow consists of flakes with an open, feathery structure and a low density. Each new fall of snow compresses and compacts the layer beneath, causing the air to be expelled and converting low density snow into higher density ice. Snow that survives one summer is known as firn and is eventually compacted to a high enough density and a thickness of about 100 m to become glacier ice. This process is known as **diagenesis** and can take many decades or even centuries.

### Valley glaciers and ice sheets

Glaciers are large, slow-moving masses of ice.

**Ice sheets** are the largest accumulations of ice, defined as extending for more than 50,000 km$^2$. There are currently only two, Antarctica and Greenland, but during the last glacial period huge ice sheets also covered much of Europe. The Antarctic ice sheet is the bigger, with a volume of about 30 million km$^3$.

**Valley glaciers** are confined by valley sides. They may be outlet glaciers from ice sheets or fed by snow and ice from one or more **corrie glaciers**. They follow the course of existing river valleys and are typically between 10 and 30 km in length

## Warm-based and cold-based glaciers

Warm-based (temperate) glaciers usually have:

- high altitude locations
- steep relief
- basal temperatures at or above pressure melting point
- rapid rates of movement, typically 20–200 m per year.

Locations such as the Andes and the Rockies experience high rates of accumulation in the winter and high rates of ablation in the summer. This makes them very dynamic, with large volumes of ice being transferred across the firn line and significant seasonal advance and retreat.

Cold-based (polar) glaciers are characterised by:

- high latitude locations
- low relief
- basal temperatures below pressure melting point and so frozen to the bedrock
- very slow rates of movement, often only a few metres a year.

In Antarctica and Greenland, summer temperatures are often below freezing and precipitation is low. This means that both accumulation and ablation are very limited and there are no great seasonal differences. The glaciers are not very dynamic and there is only limited movement.

A key difference between warm- and cold-based glaciers is their basal temperature and its relationship with the pressure melting point, as this largely determines the mechanism of movement. The pressure melting point is the temperature at which ice is on the verge of melting. At the surface this is 0°C, but within an ice mass it will be fractionally lowered by increasing pressure exerted by the overlying ice. Most temperate glaciers are at or above pressure melting point at the base and sometimes within the glacier itself. Movement is helped by the production of meltwater. However, in polar glaciers temperatures are below pressure melting point and movement is limited.

**Knowledge check 12**

How is glacier ice formed?

## Basal sliding and internal deformation

There are a number of factors that influence the movement of glaciers:

- gravity
- gradient
- thickness of the ice
- internal temperature of the ice
- glacial budget.

The sides and base of a glacier tend to move more slowly than the top and middle. This is because the ice may be frozen onto the rocks of the valley floor and sides. There may also be obstructions that create frictional resistance and slow down movement. It is also due to the accumulative effect of laminar flow in which each lower layer of ice not only moves itself, but carries the layers above with it.

### *Basal sliding*

Warm-based (temperate) glaciers mainly move by **basal sliding**. If the basal temperature is at or above pressure melting point a thin film of meltwater exists between the ice and the valley floor and so friction is reduced.

Basal sliding involves different mechanisms.

- Slippage is where the ice slides over the valley floor as the meltwater has reduced friction between the base of the glacier and the valley floor.
- Creep or regelation is when ice deforms under pressure due to obstructions on the valley floor. This enables it to spread around and over the obstruction, rather as a plastic, before re-freezing again when the pressure is reduced.
- Bed deformation is when the ice is carried by saturated bed sediments moving beneath it on gentle gradients.

### *Internal deformation*

Cold-based (polar) glaciers are unable to move by basal sliding as the basal temperature is below pressure melting point. Instead they move mainly by **internal deformation**.

Internal deformation has two elements:

- intergranular flow, when individual ice crystals re-orientate and move in relation to each other; and
- laminar flow, when there is movement of individual layers within the glacier.

When ice moves over a steep slope it is unable to deform quickly enough and so it fractures, forming crevasses. The leading ice pulls away from the ice behind which has yet to reach the steeper slope. This is **extending flow**.

When the gradient is reduced, **compressing flow** occurs as the ice thickens and the following ice pushes over the slower-moving leading ice.

**Extending flow** involves ice moving quickly down steep slopes, becoming thinner and fracturing.

**Compressing flow** occurs on gentle gradients when ice moves slowly and thickens.

# How glaciated landforms develop

Glacial landforms develop due to a variety of interconnected climatic and geomorphic processes.

**Knowledge check 13**

What is the main type of movement in
(a) valley glaciers and
(b) ice sheets?

## The influence of flows of energy and materials on geomorphic processes

There are a number of geomorphic processes that occur in glaciated landscapes.

### *Weathering*

**Weathering** happens everywhere, but different types of weathering are more or less significant in different types of landscape.

#### *Physical (mechanical) weathering*

Physical weathering breaks rock down into smaller fragments of the same material. In glaciated landscapes, it happens through the following processes.

- Freeze–thaw: water enters cracks/joints and expands by nearly 10% when it freezes. This exerts pressure on the rock causing it to split or causing pieces to break off.
- Pressure release: when overlying rocks are removed by weathering and erosion, or when overlying ice melts, the underlying rock expands and fractures parallel to the surface.

**Weathering** is the breakdown and decay of rock through exposure to the Earth's atmosphere, organisms and water.

- Frost shattering: at extremely low temperatures, water trapped in rock pores freezes and expands. This creates stress which disintegrates rock into small particles.

### Chemical weathering

Chemical weathering involves chemical reactions causing the alteration of rock minerals into different products. In glaciated landscapes this is limited as most chemical reactions are limited by low temperatures. However, some processes do occur.

- Oxidation: some minerals, especially iron, react with oxygen, either in the air or in water. The rock becomes soluble under strongly acidic conditions and the original structure is destroyed.
- Carbonation: rainwater is a weak carbonic acid. This reacts with calcium carbonate in rocks such as limestone to produce calcium bicarbonate, which is soluble.
- Solution: some minerals are soluble in water and as they dissolve they weaken the structure of a rock.
- Hydrolysis: this is a chemical reaction between rock minerals and water. Silicates combine with water producing secondary minerals such as clays.
- Hydration: water molecules added to rock minerals create new minerals of a larger volume. This happens to anhydrite, forming gypsum.

**Exam tip**

When answering exam questions about the effects of weathering, show that you understand the changes that happen to rocks. Physical weathering tends to produce smaller fragments of the same material, whereas chemical weathering produces chemically altered substances.

**Knowledge check 14**

What is the difference between weathering and erosion?

### Biological weathering

Biological weathering may consist of physical actions such as the growth of plant roots or chemical processes such as chelation by organic acids. The limited vegetation cover in many glaciated environments means that this process may be of limited importance, but does include the following.

- Tree roots: roots grow into cracks or joints in rocks and exert outward pressure, causing rock to split. Burrowing animals may have a similar effect.
- Organic acids: acid may be released during the decay of organic matter or released by algae. This acid can react with rock minerals.

## Mass movement

**Mass movement** processes may move material on slopes in glaciated landscapes, especially on steep valley sides.

- Rock fall: on slopes of 40° or more rocks may become detached by physical weathering processes. These then fall to the foot of the cliff under gravity.
- Slides: these may be linear, with movement along a straight-line slip plane, such as a fault or a bedding plane between layers of rock, or rotational, with movement taking place along a curved slip plane. Rotational slides are also known as slumps.

**Mass movement** is the downslope movement of material under gravity, without the aid of a moving force.

## Glacial processes

### Erosion

Most erosion in glaciated landscapes is carried out by moving ice in glaciers.

There are two main processes of glacial erosion. Plucking (quarrying) happens when meltwater seeps into joints in the rocks of the valley floor/sides. This then freezes and becomes attached to the glacier. As the glacier advances it pulls pieces of rock away. Abrasion occurs as a glacier moves across the surface. The debris embedded in its base/sides scours surface rocks, wearing them away.

Rates of glacial abrasion are very variable and are influenced by a number of factors.

- Presence of basal debris: pure ice is unable to carry out abrasion of solid rock.
- Debris size and shape: larger and angular debris is more effective in abrasion than fine, rounded material.
- Relative hardness of particles and bedrock: abrasion is most effective when hard, resistant rock debris at the glacier base is moved across a weak, soft bedrock.
- Ice thickness: the greater the thickness of overlying ice, the greater the pressure exerted on the basal debris and the greater the rate of abrasion.
- Basal water pressure: if basal water is under pressure, the glacier can be buoyed up therefore reducing pressure and erosion.
- Sliding of basal ice: the greater the rate of sliding, the more potential there is to erode as more debris is passing across the rock per unit of time.
- Movement of debris to the base: abrasion does not only wear away the bedrock, it also wears away the basal debris. Debris needs to be replenished if abrasion is to remain effective.
- Removal of fine debris: to maintain high rates of abrasion, **rock flour** needs to be removed by meltwater so that the larger particles can abrade the bedrock.

**Rock flour** is very fine material produced by abrasion. It gives glacial meltwater streams a milky appearance.

### *Nivation*

Nivation is a complex process, thought to include a combination of freeze–thaw weathering, solifluction, transportation by running water and chemical weathering.

### *Transportation*

Sediment can be moved within a glacier. It can come from variety of sources:

- rock fall: weathered debris falls under gravity from the exposed rock above the ice, down onto the edge of the glacier
- avalanches often contain rock debris that moves under gravity
- debris flows: in areas of high precipitation and occasional warmer periods, melting snow or ice can combine with scree, soil and mud
- aeolian deposits: fine material carried and deposited by wind, often blowing across outwash deposits
- volcanic eruptions are a source of ash and dust
- plucking: large rocks plucked from the side and base of valleys
- abrasion: fine material worn away from valley floors and sides.

Whilst being transported, the material may be classified according to its position in the glacier. Supraglacial debris is that carried on the surface of a glacier. Englacial debris is within the glacial ice and subglacial debris is embedded in the base of the glacier.

### *Deposition*

Glaciers deposit their load when their capacity to transport material is reduced. This usually occurs as a result of ablation during seasonal periods of retreat or during deglaciation. However, **till** can also be deposited during advance or when the glacier becomes overloaded with debris.

**Till** is material deposited directly by the ice.

There are two types of till. **Lodgement till** is deposited by advancing ice. Due to the downward pressure exerted by thick ice, subglacial debris may be pressed and pushed into existing valley floor material and left behind as the ice moves forward.

**Ablation till** is deposited by melting ice from glaciers that are stagnant or in retreat, either temporarily during a warm period or at the end of the glacial event.

Both types of till typically have three distinctive characteristics. Till is:

- angular or sub-angular in shape because it has been embedded in the ice and has not been subjected to further erosion processes, particularly by meltwater which would make it smooth and rounded
- unsorted: when glaciers deposit material, all sizes are deposited together
- unstratified: glacial till is dropped in mounds and ridges rather than in layers.

## The formation of distinctive landforms

### *Erosional landforms*

**Table 1** Erosional landforms in glaciated landscapes

| Name | Description | Explanation |
| --- | --- | --- |
| Corries | Armchair-shaped hollows found on upland hills or mountainsides. They have a steep back wall, an over-deepened basin and often a lip at the front, of solid rock or morainic deposits. | Nivation forms small hollows in which snow collects and accumulates, forming ice. Rotational movement of the ice under its own weight enlarges the hollow further. Plucking steepens the back wall and abrasion deepens the hollow. The thinner ice at the front is unable to erode so rapidly and a lip is left. The lip may also consist of moraine deposited by the ice as it moves out of the corrie. |
| Arêtes | Narrow, steep-sided ridges found between two corries. The ridge is often so narrow that it is described as knife-edged. | Arêtes form from glacial erosion, with the steepening of slopes and the retreat of corries that are back to back or alongside each other. |
| Pyramidal peaks | Angular, sharply pointed mountain tops. | If three or more corries develop around a hill or mountain top and their back walls retreat, the remaining mass will be steepened to form a pyramid shape. Weathering of the peak may further sharpen its shape. |
| Troughs | U-shaped valleys with relatively straight planforms. | As glaciers flow down pre-existing river valleys, they erode the sides and floor of the valley causing the shape to become deeper, wider and straighter. |
| Roches moutonnées | Asymmetrical projections of resistant rock found on the floor of glacial troughs. | As advancing ice passes resistant rock outcrops, there is localised pressure melting on the up-valley side. This area is smoothed and streamlined by abrasion. On the down-valley side pressure is reduced and meltwater re-freezes, resulting in plucking and steepening. |
| Striations | Scratches or grooves on exposed rocks in glaciated areas. | They result from abrasion by debris embedded in the base of the glacier as it passes over bare rock. Large, angular debris can produce deep grooves. |
| Ellipsoidal basins | Large, shallow basins with planforms similar to an ellipse. | Eroded by ice sheets with the surface also being lowered by the weight of the ice mass. |

**Knowledge check 15**

What is the difference between mass movement and transportation?

## *Depositional landforms*

**Table 2** Depositional landforms in glaciated landscapes

| Name | Description | Explanation |
|---|---|---|
| Terminal moraines | Ridges of till extending across a glacial trough, usually steeper on the up-valley side and tending to be crescent shaped. | They mark the position of the maximum advance of the ice and were deposited at the glacier snout. Their crescent shape is due to the position of the snout, further advance having occurred in the centre of the glacier than at the edges. |
| Lateral moraines | Ridges of till running along the edge of a glacial trough. | Material accumulates on top of a glacier, having been weathered from the exposed valley sides. As the glacier melts, this material sinks through the ice to the ground and is deposited. |
| Recessional moraines | A series of ridges running across a glacial trough, lying broadly parallel to each other and to the terminal moraine. | They form during a temporary stand-still in retreat as material carried to the snout of the glacier is deposited across the width of the trough. |
| Erratics | Individual pieces of rock, varying in size from small pebbles to large boulders, composed of a different geology to the area in which they are found. | Rock eroded, most likely by plucking, or added to the supraglacial debris by weathering and rockfall, in an area of one type of geology, and then transported by a glacier, often long distances, before being deposited during melting in an area of differing geology. |
| Drumlins | Asymmetrical, elongated hills composed of glacial till. The higher and wider **stoss**, or blunt, end faces the ice flow, while the lee side is more gently tapered. | They may be formed by lodgement of subglacial debris as it melts out of the basal ice layers, the reshaping of previously deposited material during a subsequent re-advance or the thinning of ice as it spreads out over a lowland area, reducing its ability to carry debris. |
| Till sheets | Large, thick masses of unstratified till forming extensive and relatively flat surfaces. | Large amounts of till are deposited at the end of a period of ice sheet advance during melting and retreat. |

**Exam tip**

Each of the landforms listed should be known and their development understood. Questions may ask for explanation of their formation, but examples will not be required. However, if examples are used, and aid the answer, they can be credited.

# Glaciated landforms are interrelated

Glaciated landforms are interrelated and together make up characteristic landscapes.

## Case studies

For this part of the specification, you are required to have two case studies. At least one of these case studies should be from beyond the UK. The case studies must include one landscape associated with the action of valley glaciers, and one landscape associated with the action of ice sheets.

### Landscape associated with valley glaciers

This case study could be an upland landscape glaciated in the past, such as the Lake District or the Cairngorms, or an area of active glaciation, such as the Rockies or the Himalayas.

For your case study, you should be able to illustrate the following.

- The physical factors which influence the formation of landforms within the landscape system. These are likely to include climate, geology, latitude, altitude, relief and aspect.
- The interrelationship of a range of landforms within the characteristic landscape system. These might include erosional landforms, such as corries, arêtes, pyramidal peaks, troughs, roches moutonnées and striations, as well as depositional landforms, such as various moraines, erratics and drumlins.
- How and why the landscape system changes over time. This should include short-term changes, such as rock fall adding debris to a glacier, medium-term changes, such as seasonal changes in the mass balance of a glacier, and long-term changes, such as the erosion of troughs over millennia.

You should be able to explain:

- how the physical factors have influenced the geomorphic processes
- how the geomorphic processes have led to the development of the landforms
- how the landforms interrelate with each other
- how and why the landscape system changes over time
- how and why energy and material are transferred through the glaciated landscape system.

To do this successfully, you should know:

- the names, locations and approximate scale of a range of landforms
- the rock types forming the geology of the area, and their relative resistance to geomorphic processes
- the climatic conditions and dates when the processes were acting upon the landscape.

### Landscape associated with ice sheets

This case study could be a landscape glaciated in the past, such as the Laurentian (Canadian) Shield, or an area of active glaciation, such as Greenland. However, it can be difficult to identify landscape features if the area is still under ice sheet cover.

For your case study, you should be able to illustrate the following.

- The physical factors which influence the formation of landforms within the landscape system. These are likely to include climate, geology, latitude, altitude, relief and aspect.
- The interrelationship of a range of landforms within the characteristic landscape system. These might include erosional landforms such as ellipsoidal basins and striations, as well as depositional landforms such as various moraines, drumlins, till sheets and proglacial lakes.
- How and why the landscape system changes over time. This should include short-term changes, such as weathering adding debris to an ice sheet, medium-term changes, such as seasonal deposition of till, and long-term changes, such as the erosion of basins over millennia.

You should be able to explain:
- how the physical factors have influenced the geomorphic processes
- how the geomorphic processes have led to the development of the landforms
- how the landforms interrelate with each other
- how and why the landscape system changes over time
- how and why energy and material are transferred through the glaciated landscape system.

To do this successfully, you should know:
- the names, locations and approximate scale of a range of landforms
- the rock types forming the geology of the area, and their relative resistance to geomorphic processes
- the climatic conditions and dates when the processes were acting upon the landscape.

**Exam tip**

Particularly for essay writing, it would be helpful to be able to draw a sketch map of the area with the key features labelled.

# Glacial landforms evolve as climate changes

How do glacial landforms evolve over time as climate changes?

## Glacio-fluvial landforms

Glacio-fluvial landforms exist as a result of climate change at the end of glacial periods. Glacio-fluvial landforms are produced by meltwater from glaciers. They can include both erosional and depositional landforms, although only depositional ones are required in this specification. Meltwater is released from glaciers during short, seasonal periods of melting, but mainly during deglaciation. This is when most glacio-fluvial landforms visible in the present landscape were formed. Those produced during seasonal melting and retreat are often modified or degraded during periods of subsequent re-advance.

### *Post-glacial climate change and its effect on geomorphic processes*

Glacial periods end when global temperatures rise. They are followed by shorter inter-glacials lasting from 10,000 to 15,000 years. In the post-glacial period temperatures often increase gradually, with many fluctuations, as part of a general warming trend.

#### *Glacio-fluvial deposition*

Glacio-fluvial streams and rivers deposit a distinctive sediment type known as **outwash**.

**Outwash** is material deposited by meltwater streams and rivers.

In contrast to till, deposited directly by glacial ice, outwash tends to be:
- smaller, as meltwater streams typically have less energy than glaciers and so only carry finer material
- smoother and more rounded because of contact with water and attrition
- sorted, with the largest material found furthest up the valley and progressively finer material with distance down the valley; this sequential deposition results from the progressive loss of energy
- stratified vertically, with distinctive seasonal and annual layers of sediment accumulation.

These processes influence landforms of glacio-fluvial deposition.

## Landforms of glacio-fluvial deposition

### Kames

A **kame** is a hill or hummock composed of stratified sand and gravel laid down by glacial meltwater. There are two types of kame.

1. **Delta kames** form in different ways. Some are formed by englacial streams emerging at the snout of the glacier. They lose energy at the base of the glacier and deposit their load. Others are the result of supraglacial streams depositing material on entering ice marginal lakes, losing energy as they enter the static body of water. Some also form as debris-filled crevasses collapse during a period of retreat.
2. **Kame terraces** are ridges of material running along the edge of the valley floor. Supraglacial streams on the edge of the glacier pick up and carry lateral moraine which is later deposited on the valley floor as the glacier retreats. The streams form due to the melting of ice warmed in contact with the valley sides, as a result of friction and the heat retaining properties of the valley-side slopes. Although they may look similar to lateral moraines, they are composed of outwash deposits that are more rounded and sorted.

### Eskers

An **esker** is a long, sinuous ridge of stratified sand and gravel deposited by glacial meltwater. Material is deposited in subglacial tunnels as the supply of meltwater decreases at the end of the glacial period. Subglacial streams may carry huge amounts of debris under pressure in confined tunnels at the base of the glacier. Deposition occurs when the pressure is released as meltwater emerges at the glacier snout. As the glacier snout retreats, the point of deposition gradually moves backwards. Some eskers are beaded — the ridge showing significant variations in height and width — with the beads of greater size due to periods when the rate of retreat slowed or because of seasonal variations in deposition rate.

### Outwash plains

An **outwash plain** (also known as a sandur) is a flat expanse of sediment in the proglacial area. Meltwater streams gradually lose energy as they enter lowland areas beyond the ice front and they deposit their load. The largest material is deposited nearest the ice front and the finest further away. Outwash plains are typically drained by **braided streams**.

**Braided streams** are rivers subdivided by numerous small islands (eyots) and channels.

Debris-laden braided streams lose water at the end of the melting period and so can carry less material. Material is deposited in the channel forming a mid-channel bar and causing it to divide. As energy is lost, the coarsest particles in the load are deposited first, and finer material is then added to the bar, increasing its size. When exposed at times of low discharge, channel bars can be colonised by vegetation and become more permanent features. The river divides around the island and then re-joins. Unvegetated bars lack stability and often move, forming and re-forming with successive flood or high discharge events.

## Subsequent modification of these landforms

As with till deposits, outwash deposits are often difficult to identify in the field. Again, repeated advance and retreat modify and alter the appearance of landforms which are also subject to weathering, erosion and colonisation by vegetation in post-glacial periods.

Present-day and future increases in global temperature mean higher rates of melting and retreat of glaciers. This produces more meltwater and leads to greater accumulation of outwash material in the proglacial zone. Kames and eskers will be exposed in greater numbers and of greater length during this continued retreat.

As temperatures increase, so does the growing season for vegetation. Exposed outwash material tends to become colonised over time, first by mosses and lichens and then by grasses, flowering plants and shrubs.

**Knowledge check 16**

What are the three main differences between till and outwash material?

## Periglacial landform development

Periglacial landforms exist as a result of climate change before and/or after glacial periods. Periglacial landscapes have traditionally been referred to as being 'at or near ice sheets'. However, they are more accurately defined as areas with:

- permafrost: perennially frozen ground overlain by an active layer
- seasonal temperature variations: above zero in summer, albeit for a short period
- freeze–thaw cycles, which dominate geomorphic processes.

Periglacial landscapes make up 25% of the Earth's land surface and it is estimated that another 25% has experienced periglacial conditions in the past. Periglacial conditions dominated southern England during the most recent glacial period. Today, they are found in high latitude areas (e.g. Alaska), continental interiors (e.g. Siberia) and in high mountains at lower latitudes (e.g. the Plateau of Tibet).

### Climate change and its effect on geomorphic processes

Just before and just after glacial periods, temperatures would not have been as low as during the glacial period itself, with mean temperatures often just below 0°C. There would have been many more fluctuations in temperature above and below 0°C, both from day to day and **diurnally**.

**Diurnal** temperature changes are those occurring over a 24-hour period, between day and night.

#### Freeze–thaw weathering

Freeze–thaw weathering is a dominant process in periglacial environments. Water enters cracks/joints and expands by nearly 10% when it freezes. This exerts pressure on the rock causing it to split or pieces to break off. This is due to fluctuations in temperature around freezing. The greater the number of these freeze–thaw cycles, the more effective the process is at breaking down rock. In truly glacial climates (e.g. Antarctica) temperatures are invariably below freezing throughout the year.

#### Frost heave

Frost heave is a sub-surface process that leads to a vertical sorting of material in the active layer. Stones within fine material heat up and cool down faster than their surroundings as they have a lower specific heat capacity. As temperatures fall, water beneath the stones freezes and expands, pushing the stones upwards to the surface.

#### Ground ice

The development of **ground ice** is also an important process. During summer melting periods, water percolates into the sub-surface geology where it accumulates below the water table. During the sub-zero winter months this water freezes and expands by between 9 and 10% of its volume. As this expansion occurs, so the ground surface is pushed upwards, as it is unable to extend downwards into the permafrost below.

These processes influence periglacial landforms.

**Exam tip**

Make sure you are able to distinguish between freeze–thaw weathering (on the surface) and frost heave (below the surface).

## Periglacial landforms

The landforms of periglacial areas are very varied. Periglacial landforms are a feature of current periglacial landscapes, but they are also fossil or relict features, widespread in areas that were previously periglacial.

### Patterned ground

**Patterned ground** is the collective term for a number of fairly small-scale features of periglacial landscapes. As a result of frost-heave, large stones eventually reach the surface and the ground surface is domed. The stones then move radially, under gravity, down each domed surface to form a network of **stone polygons**, typically 1–2 m in diameter. On slope angles of 3–50°, the larger stones move greater distances downslope and the polygons become elongated into stone garlands. On slopes of 60° and over, the polygons lose their shape and stone stripes develop.

**Knowledge check 17**

What is the key factor determining whether patterned ground consists of stone polygons, stone garlands or stone stripes?

### Pingos

**Pingos** are rounded ice-cored hills that can be as much as 90 m in height and 800 m in diameter. They grow at rates of a couple of cm/year. There are two types of pingo.

1. **Open-system pingos** form in valley bottoms where water from the surrounding slopes collects under gravity, freezes and expands under pressure. The overlying surface material is forced to dome upwards.
2. **Closed-system pingos** develop beneath lake beds where the supply of water is from the immediate local area. As permafrost grows during cold periods, groundwater beneath a lake is trapped by the permafrost below and the frozen lake above. The saturated **talik** is compressed by the expanding ice around it and is under hydrostatic pressure. When the talik itself eventually freezes it forces up the overlying sediments.

**Talik** is a pocket of unfrozen ground within an area of permafrost.

## Subsequent modification of these landforms

Patterned ground is a relatively minor and small-scale feature. As temperatures rise at the end of the periglacial period, patterned ground is often colonised by vegetation making it hard to find and identify. Over time, mass movement by creep also degrades the frost-heaved domes, making the landform less obvious.

Pingos collapse when temperatures rise and the ice core thaws. When this happens the top of the dome collapses leaving a rampart surrounding a circular depression called an **ognip**. Over time, the rampart itself becomes degraded by geomorphic processes making the ognip even less distinctive.

**Knowledge check 18**

What is the difference between a pingo and an ognip?

**Exam tip**

When discussing the modification of landforms, try to give an indication of the timescale involved.

# Human activity changes glaciated and periglacial landscapes

How does human activity cause change within glaciated and periglacial landscape systems?

Many glaciated and periglacial landscapes have opportunities for human activity. This includes the presence of raw materials, attractions for tourism and the potential for hydro-electric power. The socio-economic benefits of taking these opportunities can exceed the costs of overcoming the challenges involved. However, human activity on any significant scale can have major impacts on the often delicately balanced landscape systems in these environments.

## Human activity changes periglacial landscape systems

### Case study

For this part of the specification you are required to study a periglacial landscape that is being used by people. This might be related to resource extraction, such as the oil industry in Alaska, or the development of the tourist industry in places such as Northern Canada.

For your case study, you should be able to illustrate the following.

- The human activity taking place and the reasons for it taking place, such as resource availability for extraction, increasing demand and desire for energy security.
- The impacts on processes and flows of material and/or energy, through the periglacial system, such as heat produced by construction of infrastructure and pollution disrupting food webs.
- The effect of these impacts in changing periglacial landforms, such as thawing of permafrost and formation of **alases**.
- The consequence of these changes on the landscape, such as development of **thermokarst**.

You should be able to explain:

- why the human activity is taking place in the case study location
- how the human activity has affected the flows of material and/or energy in the landscape system
- how the landforms in the area are affected by the human activity and the changes it has caused to the geomorphic processes
- how the changes to the landforms have affected the landscape of the area.

To do this successfully, you should know:

- the locations and types of human activity involved
- some data as evidence of the scale and characteristics of the human activity
- the names and locations of landforms affected by the human activity
- the location and scale of the changes caused to the landscape.

## Human activity changes glaciated landscape systems

### Case study

For this part of the specification you are required to study a glaciated landscape that is being used by people. This might be related to the energy industry, such as hydro-electric power in Switzerland, or the development of the tourist industry in places such as the Lake District.

For your case study, you should be able to illustrate the following.

- The human activity taking place and the reasons for it taking place, such as dam and power station construction due to water availability, topography and increasing energy demand.
- The impacts on processes and flows of material and/or energy through the glacial system, such as the trapping of sediment behind dams and increased rates of scouring.
- The effect of these impacts in changing glacial landforms, such as increased erosion in the river channel downstream of a dam, drying up of river channels and the contraction in channel size.
- The consequence of these changes on the landscape, such as changes to the sediment flow into downstream lakes and changes to the overall drainage pattern.

| You should be able to explain: | To do this successfully, you should know: |
|---|---|
| ■ why the human activity is taking place in the case study location<br>■ how the human activity has affected the flows of material and/or energy in the landscape system<br>■ how the landforms in the area are affected by the human activity and the changes it has caused to the geomorphic processes<br>■ how the changes to the landforms have affected the landscape of the area. | ■ the locations and types of human activity involved<br>■ some data as evidence of the scale and characteristics of the human activity<br>■ the names and locations of landforms affected by the human activity<br>■ the location and scale of the changes caused to the landscape. |

**Exam tip**

Particularly for essay writing, it would be helpful to be able to draw a sketch map of the area with the key features labelled.

## Summary

- Glaciated landscape systems consist of flows of energy and material, and glacier mass balance.
- Glaciated landscape systems are influenced by a range of factors: climate, geology, latitude, altitude, relief and aspect.
- Glacier ice is formed by diagenesis, when each new fall of snow compresses and compacts the layer beneath, and there are different types of glacier: valley, ice sheet, warm-based and cold-based.
- Glacier ice moves by basal sliding and internal deformation.
- Flows of energy and materials influence geomorphic processes.
- Distinctive landforms are formed by erosion and by deposition.
- Case studies are required of one landscape associated with the action of valley glaciers and one with the action of ice sheets.
- Climate changes have occurred during a post-glacial period and have affected glacio-fluvial processes.
- Glacio-fluvial landforms are formed by meltwater from glaciers and can include both erosional and depositional landforms.
- These landforms are modified by processes associated with present and future climate changes.
- Climate changes have occurred during a post-glacial period and have affected periglacial processes.
- Periglacial landforms are formed as a result of climate change before and/or after glacial periods.
- These landforms are modified by processes associated with present and future climate changes.
- Case studies are required of one periglacial landscape that is being used by people and one glaciated landscape that is being used by people.

# Dryland landscapes

## Dryland landscapes as systems

Dryland landscapes can be viewed as systems. A **system** is a set of interrelated elements comprising components (stores) and processes (links) that are connected together to form a working unit or unified whole. Dryland landscape systems store and transfer energy and material.

### The components of open systems

Dryland landscape systems are open systems. This means that energy and matter can be transferred from neighbouring systems as an input. It can also be transferred to neighbouring systems as an output.

- **Inputs** include kinetic, thermal and potential energy; and material from **aeolian** deposition, weathering and mass movement.
- **Outputs** include evaporation and aeolian erosion.
- **Throughputs** consist of stores, such as sediment sand dunes, and flows (transfers), such as aeolian transportation.

**Aeolian** refers to the action of wind.

**Exam tip**

It is important to be able to distinguish between the components of a system, but also to understand how the components are linked to each other.

**Knowledge check 19**

What are the three different forms of energy in dryland landscape systems?

### System feedback in dryland landscapes

If a dryland system's inputs and outputs are equal, a state of equilibrium exists. This could happen when the rate at which sand is being added to a dune equals the rate at which it is being lost from the dune, and so the dune will stay the same size. If something happens to upset this equilibrium, the system changes in order to restore the equilibrium. This is known as dynamic equilibrium, as the system produces its own response to the disturbance. This is an example of negative feedback.

**Exam tip**

The dynamic element of system feedback is not the change that occurs to disturb the equilibrium, but the response of the system to restore it. Make that clear in exam answers.

### Aridity index

The aridity index is defined by the United Nations Environment Programme (UNEP) as the ratio between mean annual precipitation (P) and mean annual potential evapotranspiration (PET): the amount of water that would be lost from water-saturated soil by plant transpiration and direct evaporation from the ground.

The annual precipitation total can also be used as a guide to aridity, but this clearly does not take into account the loss of moisture to evaporation and transpiration.

The UNEP recognises four categories of aridity, as set out in Table 3.

**Table 3** UNEP categories of aridity

| | Aridity index (P:PET) | Mean annual precipitation (mm) |
|---|---|---|
| Hyper-arid | <0.05 | <100 |
| Arid | 0.05–0.2 | 100–300 |
| Semi-arid | 0.2–0.5 | 300–600 |
| Dry sub-humid | 0.5–0.65 | 600–700 |

**Exam tip**

It can be worth explaining that the actual amount of evapotranspiration may be less than the potential amount if there is insufficient water available.

**Knowledge check 20**

How is the aridity index calculated?

# Dryland landscape systems are influenced by physical factors

A range of physical factors influence processes which shape the dryland landscape. They vary in their importance and influence spatially (from place to place) and temporally (over time). These factors can also be interrelated as one factor may influence another.

## Climate

Wind is a moving force and as such is able to carry out erosion, transportation and deposition. These aeolian processes contribute to the shaping of dryland landscapes, particularly acting upon particles of sand.

Precipitation is a key factor in influencing **fluvial** processes carried out by rivers.

Dryland rivers may be **ephemeral** or **exogenic**.

The seasonal pattern of precipitation can be very variable, especially in semi-arid and dry sub-humid environments. A useful measure of this is the rainfall variability index:

$$\text{Variability (\%)} = \frac{\text{mean deviation from the average}}{\text{the average}} \times 100$$

In semi-arid regions this is about 30% and in arid areas 60%.

Temperature is also a significant factor and temperature ranges are often large. These large ranges are both **diurnal** and annual.

During the day, incoming solar radiation is not impeded by cloud cover and ground temperatures rise rapidly. At night, the clear skies allow long-wave terrestrial radiation to pass through the lower atmosphere easily and so temperatures fall rapidly. Diurnal ranges of 15–20°C are common. Locations closer to the sea typically have smaller temperature ranges as water heats up and cools down more slowly than land.

The highest air temperatures in the shade are typically 45–55°C, although temperatures in exposed locations can reach 80°C. Evapotranspiration rates are very high because of the high temperatures. There can be huge differences between precipitation totals and potential evapotranspiration figures.

## Geology

**Lithology** refers to the physical and chemical composition of rocks. Some rock types have a weak lithology, with little resistance to erosion, weathering and mass movement. This is because the bonds between the particles that make up the rock are quite weak, as in clay. Others, such as basalt, made of dense interlocking crystals, are very resistant. Some, such as chalk, are largely composed of calcium carbonate, and so soluble in weak acids making them prone to chemical weathering by carbonation.

**Structure** concerns the properties of individual rock types such as jointing, bedding and faulting which affect the permeability of rocks. In porous rocks, such as chalk, tiny air spaces separate the mineral particles. These pores can absorb and store water, known as primary permeability. Carboniferous limestone is also permeable, because of its many joints. This is known as secondary permeability.

**Fluvial** relates to the action of flowing water in channels.

**Ephemeral** rivers only flow for part of the year, typically during a wetter season and after storms.

**Exogenic** rivers have a source of water that lies outside of the dryland area. This may be in a wetter climatic region or, as is often the case, from an area of high relief which experiences high precipitation and/or snowmelt.

**Diurnal** temperature range is the difference between daily maximum and minimum.

**Exam tip**

Make sure you can comment on the seasonal variations in the climatic factors.

**Exam tip**

When explaining the influence of geology on dryland landscapes, it can be useful to refer to how the shape of the landform, such as the profile of a pedestal rock, is affected by geology.

**Knowledge check 21**

Why is chalk classified as a porous rock?

## Latitude and altitude

There is an inverse relationship between temperature and latitude, with locations at low latitudes having higher temperatures than those at high latitudes. However, each location is unique in terms of the influence of local factors such as distance from the sea, ocean currents, winds and altitude. These factors influence both temperature and precipitation and so climatic characteristics are very variable in different dryland locations.

Latitude is the key factor in determining annual temperature range, with tropical locations being less than 10°C, whilst in higher latitudes it may exceed 30°C.

Diurnal ranges are greater in areas of high altitude where the atmosphere is thinner.

## Relief and aspect

Although latitude and altitude are the major controls on climate, relief and **aspect** have an impact on microclimate.

The steeper the relief of the landscape, the greater the resultant force of gravity and the more energy a river will have to flow. Relief barriers may also create a **rainshadow** effect which can result in very low rainfall totals.

The direction a slope faces will influence local temperature; if northern hemisphere slopes face south, they will receive far more direct solar radiation and vice versa. Some slope aspects will face the prevailing wind direction, whilst others may be sheltered from it.

**Aspect** is the direction a slope faces.

**Rainshadow** effect occurs when relief barriers such as hills and mountains force moist air to rise, cool, condense and form clouds and rain. The resulting dry air warms as it sinks on the leeward side of the relief barrier, creating dry conditions.

## Availability of sediment

Sediment tends to be readily available in dryland landscapes due to the lack of vegetation cover which would otherwise stabilise loose surface sediment due to the binding effect of root systems.

Sediment is needed for erosion, and transported sediment accumulates during deposition. Much of the sediment available is in the form of relatively fine particles of sand, although larger particles, such as small stones, rocks and boulders are also significant elements of the sediment supply.

# Different types of dryland landscapes

The characteristics of different types of dryland landscapes are described below.

## Polar drylands

High latitude, polar locations experience very dry conditions. The cause of this is the low precipitation inputs, rather than high evapotranspiration losses, given the low temperatures they experience throughout the year and the lack of moisture in the cold air. The climate of polar regions is largely determined by the sinking air, and the consequent high pressure, of the Polar Cell operating in each hemisphere. These areas often have deep **permafrost** with a shallow **active layer** at the surface which is unstable in the summer.

In the summer, some snow patches melt producing small streams and the active layer may move by solifluction, a mass movement process. Freeze–thaw cycles are common and some weathering occurs.

**Permafrost** is perennially frozen ground.

An **active layer** is at or near the surface and above the permafrost. It thaws each summer before re-freezing the following winter.

## Mid- and low-latitude deserts

The world's great hot deserts lie at the tropics. Although these deserts are predominantly **reg**, some are **erg** deserts.

**Reg** deserts are rocky or stony, whereas **erg** deserts are sandy.

In these deserts there is little seasonal pattern to rainfall, which is low and very variable. Summer temperatures are extremely high and there is little surface drainage apart from some large exogenic rivers and typically smaller ephemeral rivers. Low-latitude deserts include coastal deserts, where the presence of cool ocean currents moderates temperatures.

**Knowledge check 22**

What is the difference between an exogenic river and an ephemeral river?

Deserts in mid-latitudes are more variable in character. Generally the higher the latitude the greater the annual temperature range, especially in continental interiors. Rainfall is fairly evenly distributed but with a tendency towards a summer maximum.

## Semi-arid environments

Semi-arid environments, with an aridity index of 0.21 to 0.50, are widespread. They include the Mediterranean region and the Sahel. There is a seasonal pattern to rainfall, and convectional rainfall occurs in summer and follows the movement of the overhead sun, while the period of low sun corresponds to the dry season. With increasing distance from the Equator the rainy season gets shorter and the dry season lengthens. A summer rainfall maximum is therefore found in most semi-arid environments.

**Knowledge check 23**

Why do semi-arid environments have a seasonal rainfall pattern?

Drought is a common and often prolonged feature of semi-arid environments. Dust storms and wildfires during the dry season, and floods during the rainy season, are also common.

# How landforms of mid- and low-latitude deserts develop

Dryland landforms develop due to a variety of inter-connected climatic and geomorphic processes.

## The influence of flows of energy and materials on geomorphic processes

There are a number of geomorphic processes that occur in dryland landscapes.

### Weathering

**Weathering** happens everywhere, but different types of weathering are more, or less, significant in different types of landscape.

**Weathering** is the breakdown and decay of rock through exposure to the Earth's atmosphere, organisms and water.

#### Physical (mechanical) weathering

Physical weathering breaks rock down into smaller fragments of the same material. In dryland landscapes, it happens through the following processes.

- Freeze–thaw: water enters cracks/joints and expands by nearly 10% when it freezes. This exerts pressure on the rock causing it to split or pieces to break off.
- Pressure release: when overlying rocks are removed by weathering and erosion, the underlying rock expands and fractures parallel to the surface.

- Salt crystallisation: evaporation of water results in the growth of salt crystals, creating internal stress and rock breakdown.
- Insolation: intense solar heating of rocks at the surface causes rock minerals to expand and contract, but at different rates. Rocks are weakened and slowly break down, especially if some moisture is present.

Water is clearly important in most of these processes. Despite their aridity, some rain is present and dew is also common in drylands.

### Chemical weathering

Chemical weathering involves chemical reactions causing the alteration of rock minerals into different products. In dryland landscapes this is limited by the lack of moisture as most chemical reactions involve water. However, some processes do occur.

- Oxidation: some minerals, especially iron, react with oxygen, either in the air or in water. The rock becomes soluble under strongly acidic conditions and the original structure is destroyed.
- Solution: some minerals are soluble in water and as they dissolve they weaken the structure of a rock.
- Hydration: water molecules added to rock minerals create new minerals of a larger volume. This happens to anhydrite, forming gypsum.

### Biological weathering

Biological weathering may consist of physical actions such as the growth of plant roots or chemical processes such as chelation by organic acids. The limited vegetation cover in many dryland environments means that this process may be of limited importance, but does include the following.

- Tree roots: roots grow into cracks or joints in rocks and exert outward pressure, causing rock to split. Burrowing animals may have a similar effect.
- Organic acids: acid may be released during the decay of organic matter or release by algae. This acid can react with rock minerals.

## Mass movement

**Mass movement** processes may move material on slopes in dryland landscapes, especially on steep valley sides.

- Rock fall: on slopes of 40° or more rocks may become detached by physical weathering processes. These then fall to the foot of the slope under gravity.
- Debris flows occur when slopes become saturated by heavy rain. Large quantities of rock fragments, mud and soil move at speeds of up to 50 km/hour. Even on gentle slopes, they can travel long distances.

## Aeolian processes

### Erosion

There are three main processes of aeolian erosion.

- Deflation: winds pick up and remove fine particles. This often leaves a stony surface of coarse materials known as reg.
- Corrasion is the very slow, abrasive action of wind-blown sand against rocks. The wind mainly carries sand particles close to the ground, so the abrasive effect is limited in height.

**Exam tip**

When answering exam questions about the effects of weathering, show that you understand the changes that happen to rocks. Physical weathering tends to produce smaller fragments of the same material, whereas chemical weathering produces chemically altered substances.

**Knowledge check 24**

What is the difference between weathering and erosion?

**Mass movement** is the downslope movement of material under gravity, without the aid of a moving force.

- Attrition takes place as grains of sand carried by the wind collide with each other and impact solid rock outcrops. As a result sand grains become smaller and rounder.

### *Transportation*

Wind transports particles of sediment in three ways.

- Creep: when sand grains slide and roll across the surface due to frictional drag.
- Saltation: the downwind skipping motion of sand grains which is confined to within 1 or 2 m of the surface.
- Suspension: small particles are picked up by the wind and transported in the moving air, often vast distances and even beyond the dryland area.

### *Deposition*

Wind deposits its sediment when its capacity to transport material is reduced. This usually occurs as a result of winds subsiding or slowing down due to surface friction. Once deposited, sand attracts further deposition. This is because saltating grains are less able to rebound off soft sand compared with hard rocky surfaces. As winds gradually lose energy, they deposit the largest particles they are carrying first, and so on in a sequential pattern.

## *Fluvial processes*

Although drylands typically lack water, fluvial processes are still influential on the landscape. This is due to the presence of ephemeral and exogenic rivers, as well as the influence of past, wetter climatic periods.

### *Erosion*

Dryland rivers erode in similar ways to rivers in any other environment.

- Corrasion: the river's load is rubbed against the channel bed and banks wearing them away. Dryland rivers often have high sediment load due to the lack of vegetation roots to hold surface material in place.
- Hydraulic action: turbulent flow in rivers forces air and water into cracks and crevices in the bed/banks, forcing them to widen and break up.
- Attrition: the river's load collides with itself and with the bed and banks, wearing it away and causing particles to become smaller and more rounded.

### *Transportation*

There are four processes of fluvial transportation.

- Traction: when large particles slide and roll across the channel bed due to frictional drag.
- Saltation: the downstream skipping motion of small stones which are too heavy to be held in suspension.
- Suspension: fine particles are picked up by the flow and transported in the moving water.
- Solution: soluble sediments may be dissolved and carried invisibly in the flow.

### *Deposition*

Rivers deposit when they lose energy. This can be due to a reduction in velocity and/or volume. In drylands, the high evaporation rates and irregular rainfall mean that

**Knowledge check 25**

What is the difference between mass movement and transportation?

volume loss is quite common. Velocity loss typically occurs when there is a significant decrease in gradient, perhaps as a river leaves an upland area and reaches a lowland environment. As with aeolian deposition, fluvial deposition is sequential, with the largest particles being deposited first, and so on.

**Knowledge check 26**

Why do dryland rivers often contain lots of sediment?

## The formation of distinctive landforms

### *Erosional landforms*

**Table 4** Erosional landforms in dryland landscapes

| Name | Description | Explanation |
| --- | --- | --- |
| Wadis | Stream and river channels which are dry for most of the time and when dry have lots of debris on the channel bed. | Intense rainfall and a lack of vegetation means rivers have high energy and erosive capability to form channels, even when only briefly occupied. Debris is deposited when the rivers dry up. |
| Canyons | Narrow river valleys with near vertical sides, cut into solid rock. | Vertical erosion by high discharge, often exogenic, rivers with large sediment loads. The lack of weathering and mass movement keeps the sides steep, although geology is also a factor. |
| Pedestal rocks | Isolated, mushroom-shaped rocks with a narrow base and wider, rounded top. | Undercutting is caused by saltating sand grains (1–2m above the surface). The top may be protected from erosion by a crust of minerals left after evaporation. |
| Ventifacts | Small rocks with one or more flat faces, separated from uneven faces by sharp edges. | Flat faces are abraded by persistent winds from one direction. If winds are variable, or the stone moves, more than one face may become abraded. |
| Desert pavements | Tightly-packed, coarse-grained particles cover the surface with finer materials beneath. | Deflation lowers the ground surface through the selective removal of fine-grained particles by the wind, leaving larger particles behind which are too heavy to be picked up. These then protect the finer materials beneath from deflation. |

### *Depositional landforms*

Distinctive depositional landforms can be formed by sediment, transported and deposited by wind or water.

There are various types of sand dune, but they all need an adequate supply of sand, and winds strong and persistent enough to transport it. A typical dune has a windward slope of 10–15°, a sharp crest and a steeper leeward slope (slip face) of 30–35°. The slip face stands at the angle of repose, that is the maximum angle at which loose sand is stable. Creep and saltation transport sand up the windward slope. As sand accumulates on the crest it eventually exceeds the angle of repose, causing miniature avalanches down the slope-face which restore equilibrium. In this way dunes advance in the direction of the prevailing wind.

**Table 5** Depositional landforms in dryland landscapes

| Name | Description | Explanation |
|---|---|---|
| Barchans | Crescentic dunes, with two horns facing downwind. | Form where winds blow predominantly from one direction. The horns move faster than the main body of the dune as they are lower and offer less wind resistance. |
| Linear dunes | Straight or slightly curved dunes with steep slip faces on alternate sides. | Form parallel to the prevailing wind direction which ensures dunes remain straight. |
| Star dunes | Pyramidal dunes with slip faces on three or more arms that radiate from a dome-like summit. | Form in areas where the wind is multidirectional. Sand tends to become piled up to significant heights. |
| Alluvial fans | Cones of sediment found at the foot of steep slopes, often along mountain fronts, with a concave profile crossed by many small channels. | Where a river channel gradient changes abruptly, as a river leaves a steep upland area, the sudden loss of energy, coupled with a huge sediment load, results in deposition. The main channel may split into many smaller ones as sediment is deposited within them. |
| Bajadas | A continuous alluvial apron along the base of mountain fronts. | When several alluvial fans develop at intervals along a mountain front they often grow until they eventually merge. |

**Knowledge check 27**

Why do some ventifacts have more than one abraded face?

**Exam tip**

Each of the landforms listed should be known and their development understood. Questions may ask for explanation of their formation, but examples will not be required. However, if examples are used, and aid the answer, they can be credited.

# Dryland landforms are interrelated

Dryland landforms are interrelated and together make up characteristic landscapes.

## Case studies

For this part of the specification, you are required to have two case studies. The case studies must include one mid-latitude desert, and one low-latitude desert.

### Mid-latitude desert

This case study could be the Colorado Plateau, Gobi, Lut or Thar Deserts.

For your case study, you should be able to illustrate the following.

- The physical factors which influence the formation of landforms within the landscape system. These are likely to include climate, geology, latitude, altitude, relief, aspect and the availability of sediment.
- The interrelationship of a range of landforms within the characteristic landscape system. These might include erosional landforms such as canyons, pediments and inselbergs.
- How and why the landscape system changes over time. This should include short-term changes, such as flash floods eroding wadis or depositing alluvial fans, medium-term changes, such as seasonal changes in the erosion of canyons, and long-term changes, such as pediment development over millennia.

You should be able to explain:

- how the physical factors have influenced the geomorphic processes
- how the geomorphic processes have led to the development of the landforms
- how the landforms interrelate with each other

- how and why the landscape system changes over time
- how and why energy and material are transferred through the dryland landscape system.

To do this successfully, you should know:

- the names, locations and approximate scale of a range of landforms
- the rock types forming the geology of the area, and their relative resistance to geomorphic processes
- the climatic conditions and dates when the processes were acting upon the landscape.

**Low-latitude desert**

This case study could be a coastal desert such as the Atacama or the Namib, or an inland desert such as the Sahara or the Arabian Desert.

For your case study, you should be able to illustrate the following.

- The physical factors which influence the formation of landforms within the landscape system. These are likely to include climate, geology, latitude, altitude, relief, aspect and the availability of sediment.
- The interrelationship of a range of landforms within the characteristic landscape system. These might include depositional landforms such as various types of dune, sand seas and erosional landforms such as yardangs and zeugens.
- How and why the landscape system changes over time. This should include short-term changes, such as surface wash after a flash flood, medium-term changes, such as the migration of dunes, and long-term changes, such as deep chemical weathering of bedrock over millennia.

You should be able to explain:

- how the physical factors have influenced the geomorphic processes
- how the geomorphic processes have led to the development of the landforms
- how the landforms interrelate with each other
- how and why the landscape system changes over time
- how and why energy and material are transferred through the dryland landscape system.

To do this successfully, you should know:

- the names, locations and approximate scale of a range of landforms
- the rock types forming the geology of the area, and their relative resistance to geomorphic processes
- the climatic conditions and dates when the processes were acting upon the landscape.

**Exam tip**

Particularly for essay writing, it would be helpful to be able to draw a sketch map of the area with the key features labelled.

# Dryland landforms evolve as climate changes

How do dryland landforms evolve over time as climate changes? Fluvial landforms can exist in dryland landscapes as a result of earlier pluvial periods.

## Fluvial landforms

Dryland landforms have been influenced by previous pluvial conditions.

### *Climate changes and the resultant pluvial conditions*

Over the past 500,000 years the climate in drylands has varied between wet and dry conditions. Many present-day landforms are relict features, formed during a period when the climate was much wetter than today. The most recent **pluvial** occurred between 10,000 to 6,000 years ago.

**Pluvials** are wetter periods in dryland environments, often lasting for thousands of years.

In the Sahara Desert at that time, rainfall averaged between 100 and 600 mm a year and lakes and permanent river system existed.

### The influence of pluvial conditions on geomorphic processes and landforms

Pluvial conditions intensified weathering and mass movement and increased the importance of fluvial action. Past pluvial episodes were associated with warmer conditions which reduced the significance of freeze–thaw weathering. However, rates of hydration and chemical weathering would have been greater.

Streams and rivers flowing from plateaux and inselbergs undercut steep slopes by lateral erosion, increasing the recession rates of inselbergs. They transported large sediment loads, abraded pediments and deposited rock debris across pediments.

Present-day **playas** and small lakes are the remnants of large lakes formed in the pluvial.

**Playas** are dried-out lake beds, often covered by extensive salt accumulations left by evaporation.

Present-day alluvial fans, bajadas and canyons owe much of their development to wetter climate conditions in the past when rivers were able to carry out greater rates of erosion, transportation and deposition.

### Subsequent modification of these landforms

As drylands have become more arid in the past 6,000 years, rates of weathering, mass movement and fluvial erosion have slowed. However, with less vegetation and soil cover, run-off may have increased in intensity. More extreme events and powerful, occasional flash floods may have partially offset lower rainfall. Meanwhile, drier conditions will have increased the effectiveness of aeolian erosion and transport.

The permanent rivers of the last pluvial have disappeared, replaced by ephemeral drainage.

Future predictions of climate change in drylands are uncertain. In hot arid environments, such as Africa and Arabia, a further decrease in rainfall is forecast; rates of landscape development linked to sub-aerial processes may well slow. However, aeolian erosion and transport, with deflation and the movement of dunes, could become more widespread. If rainfall declines, but becomes more extreme, fluvial erosion might increase. In contrast, mid-latitude drylands, such as the Great Plains of the USA, are predicted to become drier and so fluvial processes will slow and the associated landforms may degrade.

## Periglacial landforms

Dryland landscapes have been influenced by colder climatic conditions. Periglacial landforms can exist as a result of earlier colder periods.

### Climate changes in a previous time period and the resultant colder conditions

Global climate has fluctuated between colder and warmer periods many times over the past 500,000 years. The most recent Pleistocene glacial reached its maximum around 20,000 years ago. Global temperatures at that time were 3° to 5°C lower than today, and large parts of Eurasia and North America were covered by ice sheets and glaciers. However, the middle latitudes were largely free of glaciers. Extreme cold, with average air temperatures about 6°C lower than today, created periglacial conditions and the key landscape-defining feature was **permafrost**.

Some of today's drylands were periglacial at that time, including the Ethiopian Highlands and the Saharan Uplands. Between 13,000 and 10,000 years ago the climate warmed and periglacial environments receded north to higher latitudes.

**Permafrost** is perennially frozen ground.

## The influence of colder climatic conditions and geomorphic processes in shaping landforms

Periglacial environments, apart from a shallow surface layer, have permafrost. The main geomorphic processes are as follows.

- Freeze–thaw weathering, which occurs when water, confined in rock joints, pores and crevices freezes, expands and causes the mechanical breakdown of rocks.
- Frost heave, which is the upward swelling of the ground surface due to the growth of ice crystals in soils and **regolith**. On slopes, expansion is perpendicular to the surface, but during melting soil particles settle vertically. This results in a slow downslope movement.
- Solifluction, which is the gravitational flow of the saturated regolith. Where regolith lies on a layer of permafrost, it is known as gelifluction.

**Regolith** is a layer of loose rock covering underlying bedrock.

Some present-day drylands contain landforms that were produced by these processes.

**Table 6** Dryland landforms produced by past periglacial conditions

| Name | Description | Explanation |
| --- | --- | --- |
| Blockfields | Extensive accumulations of boulders. | Formed by frost weathering of massively bedded rocks. |
| Nivation hollows | Small, shallow depressions occupied for part of the year by a snow patch. | Physical weathering and frost-heave occur beneath well-established snow patches. The resulting debris is then removed by flowing meltwater. |
| Solifluction deposits (head) | Accumulations of unsorted, fine debris, often at the base of a slope. | Saturated regolith moves downslope under gravity over the permafrost layer and is deposited when energy is lost. |
| Frost-shattered debris | Accumulations of angular rocks. | Rocks, weathered by ice forming and melting, move downslope under gravity and accumulate towards the base of slopes. |

## Subsequent modification of these landforms

The periglacial landforms in dryland landscapes are mainly relict features, formed in an earlier, colder climatic period. The processes that formed them are therefore no longer occurring. Present-day climate and geomorphic processes will modify, degrade and eventually destroy these relict features.

In low latitudes, boulders forming blockfields are subjected to weathering processes such as insolation and salt weathering which slowly results in disintegration of the rock. Rock glaciers, which under periglacial conditions were moved by the action of ice, become inactive. Occasional flash floods can transport rock glacier particles into wadis.

**Knowledge check 28**

Why are relict nivation hollows often hard to find in present-day dryland landscapes?

In drier, warmer conditions solifluction and gelifluction no longer take place. As the accumulated debris dries out, it is eroded and dissected by ephemeral streams and rivers, whilst deflation removes finer particles. Landforms like nivation hollows are often buried beneath fluvial and aeolian deposited sediment.

Future predictions of climate change in drylands are uncertain. In some locations a further decrease in rainfall is forecast; rates of landscape development linked to sub-aerial processes may well slow. However, aeolian erosion and transport, with deflation and the movement of dunes, could become more widespread, burying some of the relict landforms. If rainfall declines, but becomes more extreme, fluvial erosion might increase causing them to be degraded even more rapidly.

**Exam tip**

When discussing the modification of landforms, try and give an indication of the timescale involved.

## Human activity changes dryland landscapes

How does human activity cause changes in dryland landscape systems? Many dryland landscapes have opportunities for human activity. This includes the presence of raw materials, attractions for tourism and the potential for agriculture. The socio-economic benefits of taking these opportunities can exceed the costs of overcoming the challenges involved. However, human activity on any significant scale can have major impacts on the often delicately balanced landscape systems in these environments.

**Exam tip**

Particularly for essay writing, it would be helpful to be able to draw a sketch map of the area with the key features labelled.

### Water supply issues change dryland landscapes

Water supply issues can cause change within dryland landscape systems.

**Case study**

For this part of the specification you are required to study a dryland landscape that is being used by people; this human use should be related to water supply issues. You might have studied somewhere like the Colorado Basin, the Nile Basin or the Khushab region of Pakistan.

For your case study, you should be able to illustrate the following.

- The water supply issue taking place and the reasons for it taking place, such as water shortage due to drought, over-abstraction or over-irrigation.
- The impacts of water supply on processes and flows of material and/or energy through the dryland landscape system, such as high rates of sediment being trapped behind dams, modifying rivers to distribute and store water, falling water tables or soil salinisation.
- The effect of these impacts in changing dryland landforms, such as decreased growth of wadis, ground subsidence or waterlogging.
- The consequence of these changes on the landscape, such as reducing depositional landforms (e.g. alluvial fans) or slowing pediment development.

You should be able to explain:

- why the water supply issue is taking place in the case study location
- how the issue has affected the flows of material and/or energy in the landscape system
- how the landforms in the area are affected by the issue and the changes it has caused to the geomorphic processes
- how the changes to the landforms have affected the landscape of the area.

To do this successfully, you should know:

- the location and nature of the water supply issue involved
- some data as evidence of the scale and characteristics of the issue
- the names and locations of landforms affected by the issue
- the location and scale of the changes caused to the landscape.

## Economic activity changes dryland landscapes

Economic activity can cause change within dryland landscape systems.

### Case study

For this part of the specification you are required to study a dryland landscape that is being used by people; this human use should be related to an economic activity. This might be related to the energy industry, such as the Qattara hydro-electric power project in Egypt, or the development of the tourist industry in places such as the National Parks of Utah.

For your case study, you should be able to illustrate the following.

- The economic activity taking place and the reasons for it taking place, such as the attractions of the natural landscape, wildlife and culture for tourism.
- The impacts of human use on processes and flows of material and/or energy through the dryland landscape system, such as vegetation and cryptobiotic crust damage by dune buggy use.
- The effect of these impacts in changing dryland landforms, such as higher erosion rates on dunes causing them to degrade.
- The consequence of these changes on the landscape, such as increased loess accumulation in marginal areas.

You should be able to explain:

- why the economic activity is taking place in the case study location
- how the economic activity has affected the flows of material and/or energy in the landscape system
- how the landforms in the area are affected by the economic activity and the changes it has caused to the geomorphic processes
- how the changes to the landforms have affected the landscape of the area.

To do this successfully, you should know:

- the location and type of economic activity involved
- some data as evidence of the scale and characteristics of the economic activity
- the names and locations of landforms affected by the economic activity
- the location and scale of the changes caused to the landscape.

### Summary

- Dryland landscape systems consist of flows of energy and material, and the aridity index.
- Dryland landscape systems are influenced by a range of factors: climate, geology, latitude, altitude, relief, aspect and sediment availability.
- Different types of dryland landscapes have different characteristics: polar drylands, mid- and low-latitude deserts and semi-arid environments.
- Geomorphic processes are influenced by flows of energy and materials.
- Distinctive landforms are formed by erosion and by deposition.
- Case studies are required of one mid-latitude and one low-latitude desert.
- Climate changes have occurred during a previous period and resulted in pluvial conditions.
- Landforms are shaped by pluvial conditions.
- Landforms are modified by processes associated with present and future climate changes.
- Climate changes have occurred during a previous period and resulted in colder conditions.
- Colder conditions have shaped landforms.
- Landforms are modified by processes associated with present and future climate changes.
- Case studies are required of one dryland landscape that is being used by people and has a water supply issue, and one dryland landscape that is being used by people for economic activity.

# Changing spaces; making places

## What's in a place?

Places are defined by a combination of characteristics which change over time. All places possess a number of characteristics or features which together make up a **place profile**.

The physical geography of a place can be a very significant element of its profile but human geography tends to have the greatest influence. How and why one place is different to another place is mainly due to people and the way they live their lives.

A **place profile** is a description of a place based on the combination of the characteristics of that place.

### The characteristics of a place profile

It is possible to put the characteristics into categories. Some of these categories overlap such as demography and socio-economic. This is because in the real world, there are lots of inter-connections amongst factors such as demography, socio-economic and cultural characteristics:

- natural characteristics (physical geography) such as geology, altitude, slope angle, location next to the sea or a river
- demography, for example the number of inhabitants, their ages, gender and ethnicity
- socio-economic characteristics, such as types of families (young adults, families with children, retired people), income levels, level of education achieved (GCSE and A-level, vocational qualifications e.g. City and Guilds, degree), types of employment (e.g. skilled manual, professional)
- cultural factors, such as religious groups, local traditions, local clubs and societies
- political factors, for example the type of local, regional and national government the place is part of, local groups such as resident associations and campaigning groups
- the built environment, for example the age and style of buildings and their building materials, flats and terraced, semi-detached or detached housing, building density.

There are a number of different factors which can influence a place profile.

**Exam tip**

It is important to know the facts and figures of your local place for each of the characteristics which make up its place profile. You will then be able to compare and contrast it with a different place.

**Knowledge check 29**

What are the demographic characteristics of a place?

### Past and present connections

Past and present connections influence a place profile. Places do not exist in total isolation from the rest of the world. Over the time a place has existed it has been connected to other places. These connections can be seen in movements of people or trade, for example. And these connections rarely stay the same for very long. Present-day connections, such as journey to work patterns, tend to be clearer to see. However, past connections can be seen in the ethnic composition of a place's population, an old building such as disused docks or in street names.

Through its connections a place is (or was) linked to places at different scales. These can be local-scale links to places close by, such as where students go to school or college or where people purchase everyday goods and services. Regional connections can include journeys to work or using a specialist health service in a very large hospital. Beyond that connections can be made with places nationally and internationally. People travel for holidays, news is gathered from the rest of the country and globally, and food and clothing are imported from a wide variety of places, near and far.

**Exam tip**

Scale is an important concept in geography. You should try to analyse connections in terms of the scale at which they operate.

## Shifting flows of people

Shifting flows of people help shape a place profile. As people make up much of a place's profile, changes to the people living in a place change its profile.

Migration brings people into a place but also leads to others moving away. These flows of people can change the demography of a place. Stage in **life-cycle** is an important influence on where people live and so influences place profiles.

**Life-cycle** stage describes the age and family status of a person such as young adult, married with children or retired.

One part of an urban area can receive an influx of young adults, for example students or young professionals just starting out on their careers and living away from home. Many such areas are in the inner city. At retirement many people move away to smaller settlements which may be in rural or coastal regions.

**Exam tip**

Migration is any permanent change in where a person lives. It includes moves at all scales from local to international and you should be clear in your answers at what scale the migrations you discuss are operating.

In ACs (advanced countries) there have been considerable increases in most people's personal mobility. Through increased car ownership levels and improvements in transport technology people are able to commute further than was the case in the past. People are able to live further away from where they work which changes the place profiles of where they choose to both live and work. Many smaller settlements are now within commuting range of a major city and this has brought significant changes to these places.

## Resources

Resources help shape a place profile. Places possess different natural resources which help shape their place profile. The local availability of a mineral resource, for example, can lead to the establishment of a community focused on mining that mineral. This is reflected in the human characteristics of that place. And if that resource runs out or is no longer required, the place undergoes change, although evidence of its previous profile often persists, for example in buildings.

**Exam tip**

Resources should be given a broad definition. Anything which humans can make use of can be included. For example, climate and heritage can be important resources for tourist resorts.

Increasingly, technology is a resource which strongly influences a place's profile. The rise in electronic communication, for example, has given great significance to the quality of connectivity a place has via the internet and mobile technology such as phones. There are considerable differences in levels of connectivity amongst different places, for example remote rural locations and urban centres.

When a place no longer has the advantages a resource once gave it, its profile can soon become characterised by decline. Such decline can be environmental, socio-economic, cultural or political. This leads to the need for regeneration and rebranding. If a place gains access to a new resource then an upward spiral of development can occur.

**Knowledge check 30**

Give examples of how resources influence a place profile.

## Money and investment

Money and investment help shape a place profile. Places require money to function. Flows of money and investment come and go, and this in turn brings about change to a place's profile.

In countries of all types, ACs, EDCs (emerging and developing countries) and LIDCs (low-income developing countries), governments are an important source of investment in a place. Governments operate at a variety of scales.

- Transnational: organisations operating across international borders such as the UN and EU.
- National: the organisation responsible for government within a sovereign country.
- Regional: this depends on how any one particular country is organised. In Germany there is a strong system of regional government and in the USA individual states are responsible for a great deal of 'regional' government. In the UK, counties perform many regional functions.
- Local: this also depends on local circumstances. In the UK there are local organisations such as town and parish councils who are responsible for small-scale local matters.

Government spending on characteristics of places such as infrastructure, education, health and environment can greatly influence place profiles.

Private flows of investment are also significant and these also operate over different scales. Transnational corporations (TNCs) operate internationally and move investment around depending on where in the world they can gain the greatest return. Substantial change can be brought to a place when a TNC either invests in it or leaves.

Businesses also operate nationally and regionally in similar ways to TNCs in terms of investment. Some firms have a very close connection with a particular place and help give that place a distinctive character.

When investigating the place profile of a place at the local scale, small-scale investments from local businesses can be significant. Business and industry is not all large factories or offices. The opening up or closing down of a shop, factory or office in a local community can make a big difference to a local place. The growth of **clone high streets or towns** is one example of where the loss of local investment changes a place's profile.

## Ideas

Ideas help shape a place profile. Because place profiles are so influenced by people, ideas can play a major role in making places what they are. In ACs, as both economy and society have moved to a **post-industrial** context, the **service sector** has been playing an increasing role in people's lives. The service sector, which includes both tertiary and quaternary sectors, is based on ideas. Increasingly the term **the knowledge economy** is used to describe these elements.

Some places are able to participate fully in the knowledge economy and so they develop a distinctive place profile. Cities which are in the top group of any hierarchy, be it of world cities or those within a country, tend to be actively involved in the knowledge economy. Their place profiles suggest thriving, lively places. Those places which are finding it hard to take advantage of the knowledge economy can have profiles suggesting stagnation or even decline.

A **clone high street or town** is one dominated by chain shops which can be found all over a country. Few local, independent businesses survive.

**Post-industrial** refers to when manufacturing industry no longer dominates geography, economy nor society. Most employed people are in services.

**The knowledge economy** describes the activities which gather, store and analyse knowledge e.g. high-tech manufacturing, finance, telecommunications, business services, design, education and health.

**Exam tip**

Make sure you can identify how past and present connections, shifting flows of people, resources, money and investment, and ideas interact to give places distinctive profiles.

A key aspect of a place's ability to benefit from ideas is the educational achievements of its inhabitants. Where such achievements are at high levels then people find openings in the knowledge economy and vice versa.

**Knowledge check 31**

What is meant by the term 'place profile'?

## Case studies

For this part of the specification you are required to have case studies of two contrasting place profiles. These are to be at the **local** scale such as a village or an individual urban neighbourhood.

Your case study must illustrate:

- their natural characteristics
- their demographic characteristics
- their socio-economic characteristics
- their cultural characteristics
- their political characteristics
- their built characteristics.

You should be able to describe:

- how past and present connections with other places at any or all of regional, national, international and global scales have influenced your chosen case studies
- how past and present flows of people (e.g. migrants, commuters), resources (e.g. raw materials, technology), money (e.g. investment from government and private organisations, such as TNCs) and ideas (e.g. trends in fashion) help shape the profiles of the two case studies.

Examples of the types of contrasting places suitable as case studies are:

- an inner city place compared to a suburban place
- a remote rural place compared to an accessible rural place.

One of these places could be your own local place which could be contrasted with an inner city place such as Toxteth or a village such as Lympstone.

A key source for your local study is the Census. Data can be obtained via www.neighbourhoodstatistics.gov.uk and information at various scales can be accessed. The use of Local Super Output Areas (LSOA) or wards allows a wide range of characteristics to be investigated.

In addition, local libraries and historical groups can be helpful when researching characteristics such as past connections and flows of people and resources.

# How do we understand place?

People see, experience and understand place in different ways which can change over time.

## What is meant by 'place'?

Place is a term that is used in many different ways. It can have an **objective** meaning such as the map co-ordinates of a place or its location on a Global Positioning Satellite (GPS) system. But a place is more than just a point on the Earth's surface. Places are given **subjective** 'meaning' by people. For example, if you follow a particular sports team, the home ground will have special meaning for you. You may have a favourite place for a holiday or where to meet up with friends at the weekend.

**Objective** means what is not influenced by personal feelings.

**Subjective** means what is influenced by personal feelings.

Space is different to place. It does not have subjective meaning and simply exists between places which do have meaning. For example, your home will have subjective meaning for you as will where you study. Along your journey to school or college, there may be places which have meaning for you but there might also be spaces where you have no connections. These locations do possess objective locational co-ordinates but have no meaning for you beyond that they exist. However, for other people, they will have subjective meanings.

**Knowledge check 32**

Distinguish between 'space' and 'place'.

## People perceive places in different ways

How and why do people perceive places in different ways? Each of us sees the world around us in different ways. Our individual personal characteristics influence our perceptions. These act like filters when we give meanings to places. It is important to appreciate that the ways in which we see the geography around us is not necessarily the way other people view the same locations.

It is also the case that perception is not always down to one factor. People are a combination of the factors mentioned below and so an individual's perception is influenced by several factors. However, in order to understand the influence of perception on making places, it is useful to consider them separately.

### *The influence of age*

Your perceptions change with age. The places that had special meanings for you when you were five will have different meanings to you now. People older than you will perceive places differently to you.

Difference in age does not mean that all perceptions of a place vary. Different generations can have similar meanings for the same place. For example a place of religious observance can hold the same special meaning across the ages.

Age is a major factor on people's perceptions of where they would like to live. Life-cycle stage describes the change from young adults seeking to live independently to elderly singles. At each change in stage, perceptions about the amount of room a household needs varies. Young adults need little space and often prefer to live in locations where work is close by, to save on journey to work costs, and where after-work services such as shops, sports facilities, cinemas and restaurants are available. Young married couples with a family have a changed perception about the amount of room their household needs, for example wanting a garden. They also place a higher value on factors such as access to nurseries and schools and open space such as parks. A retired person is likely to have different priorities, such as peace and quiet.

**Exam tip**

Study your local town or city to describe the location and characteristics of places where different life-cycle groups might live. Suggest reasons for places attracting different life-cycle stage groups.

### *The influence of gender*

Traditionally, males and females have perceived places very differently, in large part due to their different roles in society. Males occupied workplaces while females were focused on the home. Divisions in society along gender lines were also evident in the male domination of sporting places although the genders were more equal in entertainment spaces such as cinemas and theatres.

During the late twentieth and twenty-first centuries, greater equality between males and females has been emerging. Hand-in-hand with this evolution has come changing perceptions of places. With the transformation in ACs to largely service-based economies, many workplaces are less male dominated. A greater diversity of roles are open to females in the service sector. Some workplaces and professions, traditionally the preserve of females, are more open to males, nursing for example.

Perceiving places as 'safe' or 'unsafe' can be strongly influenced by gender, leading to a 'geography of fear or security' developing amongst some people. Some women feel barred from certain places at night as well as from public transport serving those places.

## *The influence of sexuality*

The increasing recognition and acceptance of different sexual orientations has allowed some places to develop a meaning on grounds of sexuality. This can be seen in the development of areas in some cities where lesbian, gay, bi-sexual and transgender groups (LGBT) tend to cluster. A physical sign of such meaning can be seen in facilities such as pubs or clubs advertising themselves as LGBT venues such as in the Castro District in San Francisco and the 'village' in Manchester.

In San Francisco the emergence of LGBT clustering in distinct residential locations is in part due to LGBT people feeling that they can openly express themselves in such **enclaves**. It has also allowed a degree of political power to be secured. The election of LGBT councillors can help create a stronger sense of place for the LGBT community.

An **enclave** is a small district in an urban area surrounded by people different to those living in the enclave.

There is also an economic element to this particular sense of place emerging. It is recognised that the 'pink' pound / dollar / euro brings a regenerating force into what were often run-down urban locations.

## *The influence of religion*

Spiritual meanings have been given to a wide variety of places for millennia:

- natural features such as mountains (Mount Fuji), rivers (Ganges, Nile), volcanoes (Kilauea), lakes (Titicaca)
- human-built features, such as Stonehenge, burial mounds, pyramids and buildings such as synagogues, mosques and churches.

Sometimes, as in the case of Stonehenge, although the construction survives, contemporary society does not share the religious meaning previous peoples gave a place. Nevertheless these places often retain some element of their profile which relates to the influence of religion. The same is true of many religious places in their ability to convey something of their spiritual meaning to all people who go to them, whether they believe in that meaning or not. For example, watching the sun rise at Uluru, Australia or entering a small medieval parish church at dusk, the influence of religion on some places can be recognised as being important.

## *The influence of role*

Many people have more than one role that they take on during the day. At home you are a son or daughter, perhaps a brother or sister or a grandchild. At school or college you are a student and you may have different roles, such as prefect, member of a sports team, music group or drama society. If you have a job you then take on the role of employee.

Different roles may lead to changing perceptions about places. A mother or father may perceive the city centre in a particular way when they go there to shop or for entertainment (e.g. the cinema) with their family at the weekend. However, their perceptions of the same place may be very different during the times that they are working in an office or shop in the centre. Then again, if they go out in an evening with people of their own age for recreation, they will have different perceptions of the central place.

**Exam tip**

Perception is an important influence on the meanings people give to places. You need to be able to discuss all five of the influences mentioned. Make sure you have real-world examples to support your knowledge and understanding.

## Emotional attachment to a place influences people's behaviour and activities

Emotional attachment to places follows on naturally from considering the influence of perception as it also varies from one person to another. Memory is an important influence on emotional attachment. How we experienced places in the past leaves us with emotions about them which range from positive to negative.

There are also places that, although we may not have personally visited them, nevertheless bring about an emotional response from us and so influence the meanings of those places to us. For example, mention of a concentration camp would generate negative feelings while a tropical sandy beach under clear blue skies brings very positive feelings for most people.

One particular way in which emotional attachment influences people is when they do not have a clearly defined and self-governed homeland. Some of the group may exist as a **diaspora**.

A **diaspora** is the spread of people away from their homeland.

There are several examples of a people who, for one reason or another, do not have their own territory as they would like to. The Kurds are one group whose main spatial concentration is in the region where the borders of Iran, Iraq, Syria and Turkey come together. The Kurds have long campaigned, sometimes violently, for their own independent state, Kurdistan.

## Globalisation and time–space compression

Globalisation and time–space compression can alter a sense of place, creating feelings of familiarity or a sense of dislocation.

Until comparatively recently, at scales from the global to the regional, distance measured in kilometres was a significant influence on the meaning of place. So, many places were relatively isolated from each other. With advances in transport and communication technologies, places have been brought 'closer' together, not in terms of kilometres but in terms of the relationships and connections which can exist amongst them. The term '**global village**' is used to describe the closer connections places now have. This has led to place profiles being changed.

**Time–space compression** is a set of processes leading to a 'shrinking world' caused by reductions in the relative distance between places, e.g. reduced travel time and the internet.

The key process driving this change is **time–space compression**. Because people, goods and ideas are much more mobile than they used to be, the world is being 'squeezed together'. A quick survey of your local supermarket shelves will tell you how much food is sourced from places a long way away.

**Exam tip**

Correct use of terms such as 'time–space compression' allows your exam responses to be precise and concise, which can make a significant difference under timed conditions.

When such dramatic change in geography occurs, there are often places that win and others that lose. Some places are able to take advantage of a shrinking world, while others cannot. The loss of mining or manufacturing industries in ACs to places in LIDCs or EDCs has left some places and many thousands of people without economic security. In some LIDCs and EDCs small-scale farmers have been out-competed by transnational corporations from ACs in the markets for their produce.

Major cities around the globe have tended to benefit from time–space compression as trade in goods and services flows amongst them. In particular, financial centres such as London, New York and Tokyo have emerged as very significant hubs in the global and national flow of money and investment.

But while some people gain by these changes to places, others do not, even in places thriving as a result of a shrinking world.

**Knowledge check 33**

Explain how time-space compression creates global hubs.

## Places can be represented in formal and informal ways

We give meanings to places that we have never personally experienced. Our perceptions of these places, as well as those we see for ourselves, rely on a wide variety of information.

**Exam tip**

Make sure you have examples of types of people who are winners and losers as a consequence of time-space compression. They need to come from all three groups of countries: ACs, EDCs and LIDCs.

### Representing places informally

There are a wide range of informal ways of representing places. The way most of us receive informal representations of places is through the media. Television and film are key in offering such images as they portray places visually, such as in soap operas (e.g. *EastEnders*, *Coronation Street* and *Emmerdale*). Through these productions we build up place profiles of places such as the rural region or urban location in which a programme is set. The combination of fiction with a real-world location can offer powerful images which can be used to market a place. New Zealand has made much of the filming of the *Lord of the Rings* trilogy to present a particular place profile to potential tourists.

Literature, art, photography, music, blogs and graffiti all convey informal images of places. It is important to appreciate that those involved in producing these informal place profiles are doing so through the influence of their perceptions. As such the images have to be analysed with care if you are trying to build up a place profile of an actual place. How realistic is the representation shown?

**Knowledge check 34**

Why do informal representations of place need to be interpreted with care?

### Representing places formally and statistically

More data about places is now collected, stored and analysed than has ever been the case. In many countries the most effective formal representations of places are their censuses. ACs such as the UK, USA and Australia hold regular surveys of their populations which are very accurate and reliable. Censuses are taken in other countries, such as many LIDCs which, while they might not be that accurate or reliable, nevertheless offer reasonable formal representations.

There has been a dramatic increase in the quantity and quality of **geospatial data** accessible to the general public. Many government agencies maintain websites which present formal representations of places. The Environment Agency produces maps showing the risk of flooding and maps are available from the police showing crime locations.

**Geospatial data** has clear locational information about the data at a particular point or area. Geographic Information Systems (GIS) (see page 66) which rely on digital storage and retrieval of data are often used.

Formal representations offer rational perspectives of a place profile, such as numbers of people living in a place, their ages, gender and educational qualifications. They are limited in their ability to indicate aspects of a place profile such as how people live their lives. Informal representations offer subjective views of a place. They mean different things to different people in different times and different places. They too are limited in terms of representing places.

# Economic change influences patterns of social inequality

How does economic change influence patterns of social inequality? There is an uneven spread of resources, wealth and opportunities within and between places.

## What is meant by social inequality?

All societies have inequalities amongst their populations. There will be inequalities in your school or college amongst the students and staff. Some factors which create differences amongst people seem to matter more than others. For example differences in shoe size tends to be less of an issue than factors such as skin colour, religion or age. **Social inequality** focuses on differences in access to housing, healthcare, education and employment opportunities and brings with it moral questions. **Quality of life** and **standard of living** are two terms also used when discussing inequalities.

No one factor determines social inequality, rather it is often the case that several factors interact to create differences.

Geographers identify that different social groups are unevenly spread across space. For example, there are more older people or more people of a particular ethnic group in certain locations than in others. These patterns of social inequality mean that there is also spatial inequality. Spatial inequalities exist at all scales from global to local.

**Social inequality** is the uneven distribution of opportunities and rewards for different social groups, defined by factors such as age, gender, class, sexuality, religion or ethnicity.

**Quality of life** is how far people's general well-being is met, including access to services such as health, education and leisure.

**Standard of living** is people's income and ability to afford material things such as food and water, housing and clothing and personal mobility.

## How can social inequality be measured?

Data on many of the factors indicating social inequality are collected in censuses. All ACs, many EDCs and some LIDCs hold a regular census which collects information about individuals. These data are then combined into spatial units of various sizes. It is possible therefore to see how factors such as age, education or health vary from one location to another.

In the UK, several factors are combined to give an **Index of Multiple Deprivation** including income, employment, health, education, crime rates, access to housing and services, and quality of living environment such as houses without central heating and air quality.

**Exam tip**

Being able to write about patterns at different scales gives your responses a real sense of authority, especially if you can give examples.

### *Measures of social inequality*

#### *Income*

The amount of money someone receives is a very important factor in social inequality. At the global scale the World Bank identifies the sum of US$ 1.25/day PPP as the definition of absolute **poverty**. PPP is purchasing power parity which relates the cost of obtaining a particular good or service to local costs. For example the cost of a loaf of bread is very different in the UK compared to Nepal.

**Poverty** is not having enough money to pay for a decent standard of living, that is buying food and water, clothes, housing and personal mobility.

In the UK **relative poverty** is considered a valuable measure of social inequality. It relates the poverty to the spread of income across the whole population. For the UK and throughout the EU relative poverty is set at 60% of the median household income.

**Knowledge check 35**

Why is the use of relative poverty helpful when investigating patterns of social inequality?

A widely used assessment of income inequality is the **Gini coefficient**. This is a ratio assessing the level of income inequality within a country. A Gini coefficient of 1.0 would mean that all the income in a country was in the hands of one person while a value of 0.0 indicates that everyone in a country has equal income. The UK's coefficient has been between 0.32 and 0.35 in the twenty-first century.

### Housing

Social inequality is clearly seen in the types and quality of housing people occupy and in **housing tenure**. What your home is like has a significant impact on your quality of life and standard of living.

**Housing tenure** is the system under which housing is occupied, e.g. owner-occupiers or tenants renting from a landlord.

Owner-occupiers own their accommodation outright and this is often achieved through borrowing money over many years; 25 or 30 years is common. When they have paid off the loan (mortgage), the capital value of the house or flat is theirs, which can add greatly to their wealth. The rental sector is made up of several types of renting, from private landlords, from local authorities or from housing associations. Figures for these different types of tenures can be found in the UK Census for example.

In LIDCs housing tenure is complicated, especially in slum areas where often well organised systems of landlords and tenants exist as well as owner-occupiers. The term 'squatter' should be reserved for those people living illegally on a plot of land.

### Education

Variations in **literacy** (which is a measure of the ability to read and write to a basic level) are a good indication of social inequality. Contrasts in literacy are clear at the global scale between countries. ACs and most EDCs have high levels of literacy (over 90% of people are literate) whereas many LIDCs have relatively low levels (60% or less).

However, formal education, as represented by literacy, is only part of the story. Informal education, such as learning skills from watching, and learning from family members can be significant in some contexts but cannot be measured. Learning how to tend livestock or throw a pot are valuable skills which may lead to employment and a decrease in inequality.

### Employment

Making a living is clearly important in terms of being able to afford a decent standard of living. Unemployment rates are useful in assessing relative levels of social inequality but do not tell the whole story. It is also, however, the level of income that is important. Relatively low wages, for example in ACs in rural compared to urban places, often lead to a degree of inequality. In EDCs and LIDCs vast numbers are employed in the **informal sector** which is very difficult to measure.

The **informal sector** is the part of the economy outside official recognition and record. People do not need formal qualifications to be employed in it, nor is there regulation of it.

### Healthcare

Inequality can be seen when people do not have ready access to healthcare professionals nor to facilities such as clinics and hospitals. One measure that can indicate the degree of inequality in healthcare is the number of doctors per 1,000 people. Most ACs offer good levels of healthcare across their populations but considerable differences persist from one socio-economic group to another and from one place to another.

Healthcare is also a consequence of factors such as the availability of clean water, effective sanitation and quality and quantity of diet. Social attitudes and lifestyles also have an impact on levels of social inequality. For example, some societies or particular groups within a society tolerate high consumption levels of alcohol or tobacco.

### *Access to services*

Globally there is a great difference between ACs, EDCs and LIDCs in people having access to education and healthcare but also services such as law enforcement. Three factors affect a person's ability to access services:

- the number of services
- how easy it is to get to the service, for example the quantity and quality of transport links and geographical distance
- social and economic factors such as age, gender and income.

In most countries there is a **rural-urban divide** as regards access to services, with people living in rural locations experiencing poorer service provision. Contrast in digital access can be significant although, increasingly, rural areas in ACs, EDCs and LIDCs are becoming connected via mobile technologies.

Residents in some countries, such as China, find their access to services such as the internet severely limited due to restrictions imposed by their governments.

The **Human Development Index** (HDI) is an attempt to combine economic and social indicators into a composite measure of inequality amongst the countries of the world. At the global scale it can suggest something as to how and why an individual's life chances are closely related to where we live.

**Exam tip**

You need to know the advantages and disadvantages of ways of measuring social inequality such as using income, education or health.

## Spatial patterns of social inequality

How and why do spatial patterns of social inequality vary within places? Social inequalities exist at all scales from the global to the local, between and within urban and rural places. They usually result from the interaction of several factors.

- **Income**: in all types of places social well-being is strongly influenced by a person's ability to pay for goods and services. It is not just being able to purchase the essentials which influence inequality but also **disposable income**.

  In general, the greater the income, the higher the standard of living and quality of life. Unemployment or irregular employment will increase social inequality.
- **Housing**: quality of accommodation is a key influence on social inequality. Living in cold, damp and overcrowded conditions is a recipe for ill-health. In LIDCs and some EDCs, millions live in slum housing. The authorities (governments and planners) are overwhelmed by the growth in numbers of people requiring housing and lack the resources to provide sufficient accommodation.

  In ACs housing is also a significant factor contributing to social inequality. The key aspect is one of affordability of housing, either through renting or purchase. When house price inflation far exceeds increases in incomes, those with low and or irregular incomes and those who are not homeowners to begin with, such as younger people, find themselves greatly disadvantaged. A particular divide exists in some rural places where large numbers of houses and flats have been bought by non-residents as holiday homes or to operate as holiday rentals. Young rural residents find themselves priced out of the market.

**Disposable income** is the amount left over once essentials such as food, clothing and housing have been paid for.

**Knowledge check 36**

Why is housing a good indicator of social inequality?

- **Health**: inequality is seen clearly in contrasting levels of health. Ill-health is associated with sub-standard housing, poor diet and unhealthy lifestyles. Dangerous and unhealthy working conditions affect millions of people in EDCs and LIDCs. In ACs, work and safety legislation has led to much healthier workplaces. Access to healthcare facilities is also important. Globally, great differences exist in the quality and quantity of healthcare available in LIDCs compared to most EDCs and all ACs. Within a country, access to healthcare services in rural places may be more difficult. And within large urban places, inner city areas may not have the level of health provision found in the suburbs.
- **Education** is regarded as vital by most people to raise standards of living and quality of life. One of the **Millennium Development Goals** is for every child in the world to receive at least primary education, giving them the skills to read, write and have basic numeracy. There are stark contrasts in levels of education at the global scale with LIDCs lagging behind ACs and EDCs. There is also inequality in education in gender terms with females being disadvantaged in some societies. As with other factors, there are often contrasts in access to education between rural and urban places and between inner and outer urban places. Rural and inner urban areas may not have the level of provision that other places have.

**Millennium Development Goals** are eight goals in a UN initiative launched in 2000 to focus efforts to improve people's lives in areas such as child mortality, gender equality and eradicating extreme poverty and hunger.

The idea of multiple deprivation can be useful when describing and explaining patterns of social inequality. It brings together the main issues involved in social inequality and helps understand how it is the interaction of factors that creates and sustains inequality.

**Exam tip**

It is important that you are able to describe and explain how social inequality results from the **interaction** of several factors.

# Economic change

Economic change creates opportunities as well as increasing social inequality.

## How does globalisation bring about structural economic change?

The term **globalisation** has come to represent the increasing interdependence of human life around the world. The multiple linkages and interconnections between governments, societies and individuals mean that events, decisions and activities in one place can result in significant consequences for communities and individuals in places far away.

One major economic consequence of globalisation is **global shift**. This has resulted in the structure of economies changing. ACs once dominated manufacturing whereas EDCs and LIDCs tend to be where many factories are now located.

**Global shift** describes the locational movement of manufacturing production, in particular from ACs to EDCs and LIDCs from the 1970s onwards.

Similarly, the primary sector has declined in many ACs with raw materials (e.g. mineral ores) tending to come from LIDCs and some EDCs. This **economic restructuring** has been brought about by several interacting factors.

**Economic restructuring** is the change in proportions of people working in various economic sectors, e.g. the change in ACs from secondary to tertiary employment.

Changes in transport technology, such as bulk handling and containerisation, allow vast quantities of goods to move relatively cheaply around the globe. ACs have evolved into places where most employed people work in services such as finance, health, insurance and education.

## The impacts of structural economic change

Structural economic change impacts on people and places. When mines and factories closed in ACs, job losses were an inevitable consequence. Many places had become dependent on a relatively narrow range of traditional economic activities such as heavy engineering (e.g. ship and railway building), chemicals and textiles. Such structural economic change was geographically concentrated in certain places. Worst affected were regions such as northeast England, northeast France, the Ruhr, Germany and the Great Lakes area of the USA. There were, however, opportunities for such places as environmental conditions such as air and water quality improved and, as a consequence, so did levels of health. On the other hand, substantial tracts of urban land were left derelict and contaminated. Toxins such as mercury and cadmium can have serious health risks and substantial cleaning of former industrial sites can be required before regeneration and rebuilding can take place. The 2012 Olympic site in East London, and in Cleveland (Ohio, USA) and Hamburg (Germany) are examples of where this has happened.

In the last 20 years, structural change has also been affecting the service sector. Many back-office clerical jobs and call centres have relocated to EDCs and some LIDCs. Places such as India have lower costs and an increasingly educated workforce to take advantage of these opportunities. Such offshoring also has opportunities for ACs, allowing them to focus on higher skilled activities such as research, product design and development and marketing.

Structural economic change has led to opportunities for some people and places. Firms have specialised in sectors where they have a **comparative advantage**.

**Comparative advantage** is the principle that countries or regions benefit from specialising in an economic activity in which they are relatively more efficient or skilled.

In ACs specialist high-tech manufacturing, such as aerospace, pharmaceuticals and bio-technology, has advantaged highly qualified people and some places. Some locations have built on an existing reputation for such economic activities, for example Cambridge. Other places have developed an international reputation, for example Bangalore in south central India has become a centre of excellence in IT and aerospace engineering.

## The impacts of booms and recessions

Booms and recessions impact on people and places. The economic status of a place is rarely static. Times of growth, stagnation and decline result in changes to the opportunities people have and in levels of inequality.

The **capitalist system** was the principal socio-economic system in many countries by 1900 and, apart from a few decades in the twentieth century when various countries, led by the Soviet Union and China, embraced communism and central state control, it has continued to dominate how economies and societies are organised.

The **capitalist system** is the socio-economic system in which production of goods and services takes place to generate profit.

Under this system surplus profits accumulate in the form of capital which is owned and controlled by capitalists. Prices are determined by the supply of and demand for the factors of production, land, labour and capital. A key driving force in the system is people's desire for gain and self-interest.

It has been claimed that the capitalist system goes through a series of interconnected cycles of economic change. These have become known as **Kondratieff cycles** (named after a Russian economist) and represent a roughly 50-year sequence of booms and recessions. Periods of boom are associated with technological innovation and new industries. After a while, the technology is no longer new and fewer opportunities for growth exist. During boom times, opportunities exist for people, especially those with the skills which the new technology requires.

The geography of this pattern is that technological innovation is not found everywhere. Some places attract investment and develop as centres of economic growth and rising standards of living. Why certain places become core regions is a matter of debate but factors such as education and the role of government and organisations such as political parties are identified as playing a role.

Places such as Silicon Valley, California, southeast England or Grenoble in Europe have attracted growth based on technological innovation and given high standards of living to those able to take advantage of the opportunities in these places.

Recessions impact people and places in different ways. The more educated generally cope more successfully than those with only basic qualifications. Places with a diversified economic base tend to retain a good level of wealth creation so that standards of living remain relatively high.

**Exam tip**

Although there is significant overlap with material in economics, your responses must keep a sharp focus on the **geography** of the processes and patterns you discuss. One way to do this is to set your comments firmly in the real world.

## The role of government in patterns of social inequality

Most governments in their stated aims include social justice, political cohesion and the reduction of poverty and social inequality. In many ACs, government spending on areas such as health, pensions and education amounts to very significant sums in both relative and absolute terms. In EDCs, government involvement in areas that directly impact on social inequality has been growing as their economies develop but LIDC governments largely lack the resources and organisation to be able to significantly reduce levels of social inequality.

The ageing populations of most ACs is presenting significant challenges in providing healthcare and pensions. Amongst the retired population there can be substantial inequalities depending on the level of pension they receive, whether they own their own homes and the types of jobs they had. Semi- or unskilled people often worked in more polluted and dangerous conditions so they can often have long-term health issues. They tended to live in rented accommodation and have little disposable money to save for their retirement.

**Exam tip**

Look up the latest available figures for facts such as government spending in areas such as education and pensions and the percentage of pensioners in a place and use these in your answers.

Providing services in rural areas has long been an issue of concern. In areas of low population density, governments have looked to use **key settlements** as locations for services such as education, healthcare and retailing. The availability of public transport in rural regions is another concern. These are contested matters as they depends on attitudes towards giving subsidies to allow the services to continue.

**Key settlements** are places where services (e.g. school, doctor, shop) are concentrated in rural areas so that thresholds of numbers of users are met to ensure the services survive.

Governments rely for their funding on raising money through taxation and borrowing. Both of these are finite and governments have to take decisions about spending on areas such as health, education and infrastructure which put limitations on what can be spent. The geography of social inequalities has tended to be remarkably difficult to change for a wide range of types of government.

# Social inequality impacts people and places in different ways

## Case studies

For this part of the specification, you are required to have two case studies. These are to be of contrasting places that illustrate:

- evidence of social inequality
- the range of factors that influence social inequality
- how social inequality impacts people's daily lives in different ways.

You should be able to describe evidence of social inequality for each place, such as:

- the quality of housing (e.g. detached, semi-detached, terraced, flats) and materials (e.g. brick, wood, corrugated iron sheets, plastic sheets)
- the environmental quality of the place (e.g. air, water and land quality), how densely built-up the place is and the noise levels
- crime rates (e.g. official statistics)
- the digital divide (e.g. access to telecommunications, availability and speed of digital connections).

You should be able to show how the level of social inequality is influenced by a range of factors such as:

- demographic characteristics of the people living in the place (e.g. ages, gender, ethnicity)
- social characteristics of the people (e.g. health levels; education attainment)
- economic characteristics such as income, types of jobs including factors such as full or part-time, level of personal mobility (e.g. access to public transport or car ownership level).

You should be able to relate the evidence of social inequality and factors influencing its level to the daily lives of the people living in each place.

Examples of the types of contrasting places are:

- inner city – outer suburb in an AC
- urban – rural in an AC, EDC or LIDC
- a place in an AC city – a place in an EDC or LIDC city
- accessible rural – remote rural in an AC.

**Exam tip**

You must be able to write in detail about your chosen examples. This requires you to know some facts and figures about each one, including locational information.

# Who are the players that influence economic change in places?

Places are influenced by a range of players operating at different scales.

The term **players** is the same as **stakeholders**. Both refer to individuals, groups or formal organisations who can influence, or can be influenced by, the processes of change in a place. There is a very wide range of players operating across different scales from the international to the private individual.

**Exam tip**

The specification only uses the term 'player' so this is the term which will be used in exam questions.

## The role of players in driving economic change

**Public players** include governments and these operate across the scales. The EU is a transnational government which can have direct economic influence on a place. It can issue grants to support the construction of infrastructure, such as road widening or bridge building. National governments have departments responsible for strategic planning which drives economic change, such as education and training and major transport links. Government spending on defence can lead to military bases expanding or closing, and the placing or not of orders for military equipment can bring prosperity or decline to manufacturing in a place. Local government has responsibilities for planning and can bring about economic change, for example by supporting the building of an industrial estate or the regeneration of a town centre.

There is a very wide variety of **private players**. Large-scale ones include transnational corporations who make locational decisions about their business at the global scale. The opening or closing of mines, factories or office complexes can significantly impact a place economically and socially. Locally, smaller-scale employers can also have significant effects when they expand, contract or even close.

Local communities are concerned with their immediate area. Economic change can be influenced by local attitudes towards issues such as house building on the edge of a settlement or the redevelopment of a brownfield site.

Amongst the private players are a large number of **non-governmental organisations** (NGOs), most of which have a particular focus. For example, the National Trust is concerned with the conservation of historic buildings and scenic landscapes and habitats.

**Knowledge check 37**

Why can economic change in a place often be contested?

**Exam tip**

The different scales the various players operate at needs to be clear. An example is the way national government can influence major transport change while local government deals with the redevelopment of a city centre.

## Case study

For this part of the specification you are required to have one case study of a place that has been impacted by structural economic change. This is to be at the scale of a **country** or **region**.

Your case study must illustrate:

- the structural economic change which the country or region has gone through
- the role of players in driving the changes
- the impacts on places and people living in the country or region.

You should be able to describe the following.

- The characteristics of the country or region before the economic change occurred, to include demographic, socio-economic, cultural and environmental characteristics.
- The economic changes that took place, such as changes in types of economic activities (e.g. from manufacturing to services or from agricultural to manufacturing).
- The different players involved in driving the change including their roles, such as national and/or local government, transnational corporations and international institutions (e.g. EU or World Bank).
- The impacts on the people living in the country or region of the economic change including demographic, socio-economic, cultural and environmental impacts.

Examples of the types of places suitable as case studies are:

- regions in ACs that have undergone de-industrialisation (e.g. the West Midlands, UK, the Ruhr, Germany or the Great Lakes, USA and Canada)
- regions in ACs, EDCs or LIDCs that have undergone economic growth (e.g. Exeter or Oxford regions, Bangalore in India, Pearl River Delta in China)
- EDCs or LIDCs that have undergone economic change such as industrialisation (e.g. Thailand or Vietnam).

# How are places created?

Places are created through placemaking processes. Place is produced in a variety of ways at different scales.

With the world's population growing, albeit at a slowing rate, the 7.5 billion (2016) total is expected to reach somewhere between 9 and 10 billion by 2050. The making of both urban and rural places where these people will live, work and play presents considerable challenges. This is as true of the megacities with their 10 million plus inhabitants, such as Tokyo or Mumbai, as it is of a small town in the mid-west of the USA. It is as true of the crowded rural regions, such as the island of Java, Indonesia, as it is of the remote rural region of northern Scandinavia.

## The role of governments and other organisations in placemaking

Placemaking has geography at its heart. Governments and organisations at all scales recognise this, with some 60–80% of all data including a locational component. **Geographic Information Systems (GIS)** allow data's geography to be a key element of placemaking.

**Geographic Information Systems (GIS)** are integrated computer tools for gathering, storing, processing and analysing geographic data, which is data that can be plotted on a map.

GIS makes it possible for a wide variety of data to be manipulated so that different ways of looking at a place can be considered. This in turn allows changes to places to be modelled, in the hope of anticipating issues and having appropriate measures in place before a problem arises. For example, predicting traffic flows as a result of changes to road or rail networks enables points where congestion might occur to be planned for. Retailers use GIS to predict potential customer numbers when planning the location of a new store.

Places are keen to attract inward investment so that growth can be sustained or regeneration started. Governments and a wide variety of other organisations are responsible for promoting places of all scales in order to attract such investment. Large numbers of government officials represent the UK at home and abroad. Ambassadors in virtually every country promote the interests of the UK and its citizens. Organisations such as the British Council specialise in supporting educational and cultural links with other countries. NGOs such as the National Trust organise groups of supporters in other countries to raise funds to invest in places such as the Lake District.

### *The attraction of Foreign Direct Investment (FDI)*

With the importance of transnational corporations (TNCs) to investment across the economy, governments have been keen to encourage inward investment. This has become particularly important when placemaking has had regeneration as its main aim, for example in places where de-industrialisation has been severe. Most flows of FDI come from TNCs headquartered in ACs, such as Barclays, Sony and Nestlé with over 60% of their investments going to other ACs. An increasing number of TNCs originating from EDCs and LIDCs are operating globally. For example, Tata is an Indian-based TNC which runs a vast manufacturing portfolio of companies such as steel-making and vehicle assembly in over 100 countries including many ACs.

**Exam tip**

Put together a case study of a place where FDI has been attracted, helping to change that place. Make sure you can describe the role governments and/or other organisations had in attracting the investment.

## How planners and architects make places

In the UK, every local authority makes plans for its development. Professional planners oversee the process and many different players contribute, such as elected councillors, community groups and individuals. Local neighbourhood plans are drawn up as well as a Core Strategy which guides planning. Elements of place such as industrial and housing developments, transport and amenities such as parks are included. When development is of national significance, such as a major transport route or power station, the development still has to go before the local planning system, but national strategic considerations can override local feelings.

Architecture has always been a significant factor in placemaking. Individual buildings, public spaces and the appearance of neighbourhoods is influenced by the designs architects come up with. Architecture can influence the use made of the built environment and leads to places being perceived either positively or negatively. For example, the residential high-rise blocks built in many cities in Europe between 1950 and 1980 created places which people found unappealing and led to isolation for their inhabitants. They were often poorly built, suffered from issues such as damp and were expensive to heat. In contrast, the waterfront developments of the late twentieth and early twenty-first centuries in places such as London Docklands or parts of Boston, Massachusetts, USA have generally been seen positively in terms of remaking these places.

### *The 24-hour city*

Increasingly, large urban centres are developing 24-hour rhythms of constant but different activities. Parts of the city become different places depending on the time of the day.

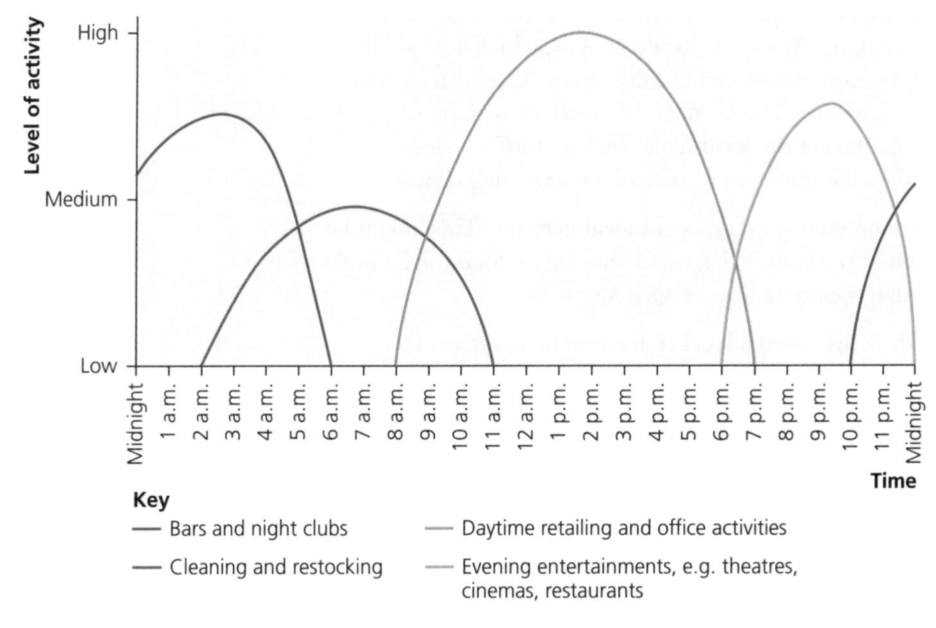


**Figure 1** Changes to an urban place through a 24-hour period

One of the reasons some cities are actively promoting themselves as 24-hour cities is rebranding of the central area to rid this location of the perception of it as deserted, threatening and unsafe.

Factors behind the transition to a 24-hour city include the following.

- Population change: rising numbers of people in central areas as run-down inner cities are regenerated and transformed. Factories, warehouses and decaying housing are renovated into flats and shops. In particular, young professional people choose to live here, who after work provide the demand for shops and leisure facilities in the evening. Berlin, London and Madrid are examples.
- Planners and architects directly support 24-hour activity through their designs. Public transport runs 24/7 and the layout of stations and design of trains, for example, have passenger safety as a high priority. Removal of restrictions on opening hours for some services such as clubs and bars allow their use through the night. Services such as gyms and hairdressers can be found open late into the night. Art exhibitions and theatres open into the early hours when particularly popular events are held.
- The rise in international tourism has given an impetus for some cities to develop their 24-hour culture, for example Las Vegas, New Orleans and Paris. Visitors provide a demand for food, drink and entertainment through into the early hours.
- Facilities open 24 hours generate journeys to work throughout the day and night. 24-hour retailing (e.g. supermarkets) means that deliveries are made at almost any time of night or day. In some cities, the authorities insist on overnight deliveries so that large delivery vehicles do not add to congestion during rush hours.

**Knowledge check 38**

What is meant by the '24-hour city'?

### How local community groups shape places

Local community groups shape the places they live in. There is often a wide range of local groups involved in placemaking. As well as local councils, which are usually dominated by the major political organisations, community groups such as residents' associations can be active in placemaking. They concern themselves with housing, community and environmental matters at the local scale, such as traffic speeds through a residential neighbourhood or the maintenance of a community centre.

Many groups are concerned with protecting an aspect of local heritage. They might be part of a larger organisation such as the National Trust or they might focus on a purely local characteristic such as a small theatre or area of open space.

These groups can be particularly active when a local characteristic is subject to possible changes. The redevelopment of a sea front or an individual site can result in local people becoming very involved in local placemaking. Increasingly, social media is used to generate and inform support and to help lobby decision-makers such as planners and developers.

## Rebranding changes a place

Rebranding changes a place through **reimaging** and **regeneration**.

### Why places rebrand

The image a place has is its brand. This is made up of objective elements such as where it is located (e.g. coastal or in a valley) but also subjective elements (e.g. is it modern or old fashioned, is it safe?). If a place has a negative image then **rebranding** involving reimaging and regeneration might be attempted.

**Rebranding** involves developments aimed at changing negative perceptions of a place, making it more attractive to investment.

**Figure 2** The process of rebranding

Rebranding and reimaging are not new. For centuries, places have reinvented themselves. The remaining buildings of ancient Rome and Athens were intended as strong statements of prosperity and power, replacing what was there before. In the nineteeth century, when towns and cities were growing rapidly and leaving behind their medieval roots, public buildings such as town halls and art galleries, and private buildings, such as railway stations, were designed to impress with their grandeur and often extravagant architecture.

There are three key elements in rebranding a place.

1. Brand artefact: the physical environment, in particular individual buildings or characteristic features such as drystone walls in upland rural regions.
2. Brand essence: the experience of people when in the place.
3. Brandscape: how the place sets itself up in relation to places it is competing against for investment.

## Strategies used to rebrand places

A range of strategies can be used to rebrand places. Most places use a combination of strategies when rebranding.

- Market-led: private investors take the lead (e.g. property developers, builders and business owners developing or rebranding retailing, restaurants, pubs and wine bars). **Gentrification** is typical of this strategy as in Islington, London or Le Marais, Paris.
- Top-down: large-scale organisations, often public sector (e.g. local authorities' planning departments), development agencies and private investors, such as insurance and pension companies. Salford Quays, Manchester and Inner Harbor, Baltimore are examples.
- Flagship development: large-scale one-off property projects using distinctive architecture. The intention is that this focused rebranding acts as a catalyst for further investment. Examples include the Millennium Stadium, Cardiff and The Waterfront, Belfast.
- Legacy: investment follows after a major sporting event such as the 2016 Olympics, Rio de Janeiro and 2014 Commonwealth Games, Glasgow.
- Events or themes: being declared the European Capital of Culture acts as a catalyst for further regeneration such as Liverpool in 2008 and Riga, Latvia in 2014.

**Gentrification** is process by which former low-income inner-city housing districts in ACs are invaded by higher-income groups and refurbished.

Whatever strategies are used, various elements tend to be involved in bringing about the transformation of a place.

**Table 6** Elements used in rebranding

| Element | How used | Examples |
|---|---|---|
| Architecture | Either to reinforce a particular heritage look or to promote a modern look. | Central Birmingham<br>Covent Garden, London<br>Pompidou Centre, Paris. |
| Heritage | Uses the history of a place to attract visitors. | Wessex (parts of Dorset, Somerset and Wiltshire)<br>Northumberland, northeast England |
| Retail | Uses consumer spending and the idea of shopping as a leisure activity when combined with eating out or visiting a cinema. Often involves flagship stores in distinctive architecture, refurbishment of existing shops and streets. | Central Birmingham<br>Dubai. |
| Art | Uses art events and permanent galleries, some with distinctive architecture, to attract visitors. | Guggenheim Museum, Bilbao, northern Spain<br>Tate Gallery, St Ives, Cornwall. |
| Sport | Uses major events to reimage and attract visitors and investment. | F1 Grand Prix Bahrain<br>Marathon run Stockholm, Sweden. |
| Food | Uses specialist and high quality food and dining to attract visitors. | Ludlow, Shropshire, England<br>Chinatown, London and San Francisco. |

## Players involved in rebranding

A range of players are involved in placemaking through rebranding. Rebranding is a diverse process with many different players involved. Amongst these, players responsible for funding are key, such as governments of various scales. The EU's **European Regional Development Fund** (ERDF) plays a major role across the EU. It helps pay for projects such as infrastructure projects (e.g. roads and railways). These can improve access into a place requiring rebranding and so help attract investment and economic activities. Making the A55 along the north Welsh coast into a dual carriageway makes this remote and declining rural region more accessible to day and short-break visitors, thereby stimulating tourism.

Corporate bodies such as financial institutions (banks, pension funds, insurance companies) and property developers often carry out actual physical regeneration of buildings. Not-for-profit groups such as National Park authorities, the National Trust, English Heritage, the Arts Council and local community organisations, such as residents' associations, also play roles in rebranding. The National Trust, for example, conserves both buildings and landscapes and attracts visitors who spend money, providing employment and a demand for goods and services.

**Exam tip**

Geography deals with many ideas such as rebranding which are 'contested', that is not everyone agrees with the idea and/or the way it is put into practice. You must be able to show you know and understand the different points of view.

## Rebranding is often a contested process

Because of the change that rebranding brings to a place, not everyone feels that it is in their best interests. There can often be tensions amongst players. Gentrification, for example, results in a different socio-economic group moving into a neighbourhood, changing the character of that place. Existing local residents can feel alienated by the changes and excluded from the new shops, wine bars and restaurants.

Different players may have different priorities in the rebranding process. Commercial organisations need to see a return on their capital investment and may focus on revenue-generating developments. Other players, such as local residents, may give a higher priority to improving local amenities, such as open space, public transport and schools and health centres.

## Making a successful place requires planning and design

### Case study

For this part of the specification you are required to have one case study of a place that has undergone rebranding. This works best at the scale of either an urban area or a rural region.

Your case study must illustrate:

- why the place needed to rebrand
- the strategy or strategies involved in the rebranding
- the role and influence of the range of players involved in the rebranding and remaking of the place
- how the rebranding altered people's perceptions of the place
- an evaluation of the relative success of the rebranding.

You should be able to describe the following.

- Characteristics of the place before rebranding occurred, to include demographic, socio-economic, cultural and environmental characteristics. You should include a summary of what the overwhelming perception of the place was before rebranding.
- The strategy or strategies that were used in the rebranding process, such as economic restructuring from primary activities (e.g. farming and forestry) and/or manufacturing to services (e.g. tourism, retailing or office based employment), the use of architecture, the use of heritage and the use of sporting and/or cultural events.
- The different players involved in driving the rebranding including their roles, such as national and or local government, transnational corporations and/or international institutions (e.g. EU, non-governmental organisations, local community groups).
- How the perception of the place has been altered as a result of the rebranding process.

You should be able to offer an assessment of the extent to which the rebranding has achieved its aims. This should also include an evaluation of how different players view the results of the rebranding, such as whether they think that their lives and/or the place are improved as a result of the rebranding.

Examples of the types of places suitable as case studies are:

- urban places that have suffered from de-industrialisation (e.g. Liverpool, Manchester, Barcelona, Baltimore, USA)
- rural regions that have experienced decline in agricultural employment and/or loss of rural industry such as forestry or mining (e.g. Snowdonia, Cornwall, Massif Central, France, Mezzogiorno, southern Italy)
- places looking to break onto a wider 'stage' (e.g. Bahrain, Qatar).

## Summary

- Places consist of several physical and human features which combine to give the place a distinctive character.
- A case study is required of two contrasting places at the local scale to show how a variety of characteristics, flows and connections interact to shape the two place profiles.
- Flows of people, goods and ideas help shape the character of a place. These change through time.
- Place is different to space as it is given meaning by people.
- People give meanings to places through their perceptions.
- Perceptions are subjective and influenced by several interacting factors such as age, gender and education.
- A person's emotional attachment to a place can influence their behaviour and activities in a place.
- Time-space compression can alter how people perceive a place.
- Places are represented in both formal and informal ways. Formal tend to be objective while informal are more likely to involve subjective feelings.
- Economic change brings about changing patterns of social inequality in places.
- Social inequalities can be recognised in factors such as housing, education, employment and healthcare.
- Social inequalities are seen in spatial differences between and within places.
- A case study is required of two contrasting places to illustrate the types of evidence of social inequality, the range of factors that influence social inequality and how social inequality impacts upon the daily lives of the inhabitants.
- Some people and places benefit from economic change but others lose out.
- Cycles of booms and recessions bring social opportunities and inequalities to people and places.
- Government, both local and national, has a major role in affecting patterns of social inequality.
- Different places can have great contrasts in the standard of living and quality of life experienced by people living there.
- Places are influenced by a variety of players (stakeholders) who operate at different scales.
- These players drive economic change which then alters the characteristics of places.
- A case study is required of a country or region that has undergone structural economic change to illustrate the characteristics of the place before the economic change, the economic changes that took place, the players driving the change and the impacts on both the people and the place.
- Places are created by a variety of placemaking processes or influences.
- Governments and organisations attract inward investment to regenerate places.
- Architects and planners use design to create places such as 24-hour cities.
- Local organisations such as community groups and heritage associations contribute to placemaking.
- Places rebrand through reimaging and regeneration using a range of strategies.
- A range of players have different roles in the rebranding process.
- Rebranding is a contested process with some people feeling marginalised.
- Rebranding needs sustained careful planning and design if it is to be successful.
- A case study is required of a place, such as a town, city or rural region, that has undergone rebranding.

# Questions & Answers

## Assessment overview

For both AS and A-level, you will be presented with a choice of three types of landscape: coastal, glaciated and drylands of which you choose **one**. Each of these three questions will have the same structure. There will be a series of sub-parts for a relatively small number of marks, and finally an essay.

At both AS and A-level, you will *have to* answer a question on Landscape Systems and Changing Spaces: Making Places. No choice is offered for Changing Spaces; Making Places and the questions have a similar structure to those in Landscape Systems.

In the AS paper, Landscape and Place, you will have to answer the Landscape Systems question, a question on Changing Spaces; Making Spaces, and a fieldwork question. The Landscape Systems essay will typically be worth 14 marks, with the other sub-parts usually ranging between 3 and 8 marks each. The total for the whole question will be 29 marks. The Changing Spaces; Making Spaces question will have a similar structure and is also worth 29 marks. In the examination you should allow 35–40 minutes to answer each of these questions worth 29 marks. The fieldwork question will be out of 24 marks, with 12 marks for an essay and the remaining 12 marks divided into a few sub-parts. In the examination you should allow about 30 minutes to answer this question. The grand total for the AS Core paper is, therefore, 82 marks and the paper is 1 hour 45 minutes long. This makes up 55% of the AS qualification.

At A-level, both the Landscape Systems and Changing Spaces; Making Places question are slightly different, although they are based on exactly the same content as for AS and are in different papers. Both questions are worth 33 marks, with 16 marks usually allocated for the essay. The other sub-parts will have questions typically worth 2–8 marks. In the examination, you will spend about 45 minutes on each question. In the other part of the Physical Systems paper, you will have to answer a question on Earth's Life Support Systems, which will have a similar structure and also be worth 33 marks. In the Human Interactions paper, you will also have to answer two questions on Global Connections, made up of shorter sub-parts and an essay, which are worth a total of 33 marks. Therefore, the grand total for each A-level paper is 66 marks and each paper is 1 hour 30 minutes long. The Physical and Human papers each account for 22% of the A-level qualification.

## About this section

The questions below are typical of the style and structure that you can expect to see in the A-level papers. Questions in the AS paper will be very similar, but with a slightly different mark allocation, as indicated above. Each question is followed by examiner comments which offer some guidance on question interpretation. Student responses are provided, with detailed examiner comments within each answer, to indicate the strengths and weaknesses of the answer and the number of marks that would be awarded. A final summary comment is also provided, giving the total mark and grade standard.

# Landscape systems

## Question 1 Coastal landscapes

**(a)** **Explain how coastal sediment can be supplied from a variety of sources.** [8 marks]

**(b)** **Study Table 1, the highest tides experienced at southern England locations (until 2012).**

**Table 1:** Highest tides experienced at southern England locations (until 2012).

| Location | Highest tide (m) |
|---|---|
| Avonmouth | 15.0 |
| Bournemouth | 3.1 |
| Dover | 8.0 |
| Ilfracombe | 10.5 |
| Newhaven | 7.7 |
| Newlyn | 6.4 |
| Plymouth | 6.4 |
| Portsmouth | 5.5 |
| Weymouth | 3.0 |

**(i)** **Calculate the median value of the data in Table 1. You must show your working.** [2 marks]

**(ii)** **Calculate the range of data values shown in Table 1. You must show your working.** [2 marks]

**(iii)** **Suggest why the highest tides vary between the different locations.** [2 marks]

**(c)** **Study Figure 1, a coastal landscape in North Devon.**

**Figure 1** A coastal landscape in North Devon

Explain how waves have developed the cliffs in this landscape. **[3 marks]**

**(d)* Assess the relative importance of past, present and future sea-level rise in influencing coastal landforms.** **[16 marks]**

Total = 33 marks

**(a)** This question requires an explanation, but the focus should be on the **variety** of sources, rather than separate explanations of each source.

**(b)** This is a skills-based question and both median and range are explicitly mentioned in the specification. Remember to show your working. In (iii) you are not expected to know anything about the specific locations; you just need to suggest appropriate, possible reasons.

**(c)** There are only 3 marks available here, so keep a clear focus on erosion; no need to mention weathering, for instance.

**(d)** The command word here is 'assess' and so you must ultimately decide which of the three sea-level rises you believe to be the most/least important. The * means that the quality of your extended response will be assessed in this question. You should use full sentences, spell and punctuate correctly and make appropriate use of technical terminology.

## Student answer

**(a)** Coastal sediment can be derived from a number of sources. Geomorphic processes transport sediment to the coast and then deposit it. This includes a variety of moving forces, including wind, rivers and waves. Wind blows fine sediment and deposits it when it loses energy due to friction from the land. Waves carry sediment and deposit it after they break and move it up the beach in the swash. Rivers bring sediment from inland and deposit it at their mouth when they lose energy when they enter the sea.

Sediment also comes from weathering and mass movement on cliffs behind beaches. Even if they are not always being directly eroded by waves, the cliff face will be weathered by processes such as salt crystallisation. Sea spray lands on the cliff face and then the water evaporates leaving the salts behind. Crystals grow in pores and cracks and as they get bigger they exert an outward force which causes the rock to break up. Weathered fragments then drop under gravity as rocks fall down onto the beach adding to the sediment that is there.

e **6/8 marks awarded.** This is a competent answer and the student has provided some focus on **variety** by commenting on the different moving forces. However, they could have noted that weathering and mass movement do not require a moving force, and so are different. Also, they haven't referred to human sources such as beach nourishment.

**(b) (i)** Median = 6.4 m. You put all the values in order of size and find the one in the middle which is the median. 3.0, 3.1, 5.5, 6.4, **6.4**, 7.7, 8.0, 10.5, 15.0

**e 2/2 marks awarded.** The student has correctly calculated the median value and has shown their working, by ranking the data and highlighting the middle (5th of 9) value.

**(ii)** 15 – 3 = 12 m

**e 2/2 marks awarded.** The student has correctly calculated the range by subtracting the lowest value from the highest, and stating the answer.

**(iii)** The height of tides varies depending on the season. The highest tides will occur in the spring and autumn months when the gravitational pull of the moon and the sun is at its highest.

**e 0/2 marks awarded.** The student has misunderstood this question, and is explaining why high tides occur at different times of year. The issue here is that the tidal heights are different in different locations. This is likely to be influenced by the shape of the coastal landscape. Open coastlines tend to have low tidal heights whereas in confined locations such as estuaries the high tides tend to be higher (and low tides lower, hence their high tidal range). This would explain the highest tides being in the Severn Estuary, Bristol Channel and Straits of Dover.

**(c)** Waves transfer energy to the coastline when they break **a**. Their force can erode cliffs by pounding and hydraulic action, when air and water are forced into cracks and joints **b** in the cliff face. Waves also erode cliffs by abrasion. When they break they hurl rocks and stones against the cliff wearing them away as rock rubs against rock. The rocks in the photograph look to be all of the same geology and so would erode at the same rate **c**. However, erosion is concentrated between high and low tide levels **d**, undercutting a notch at the base of the cliff which eventually leads to the unsupported overhang collapsing and retreating. The cliff retreats and becomes higher and steeper as it retreats.

**e 3/3 marks awarded.** The student has provided a very full, and accurate explanation. There is detailed use of specific process mechanisms **b** and there are clear links to how the cliffs develop due to undercutting, collapse and retreat **d**. They have also made an explicit link to the cliffs in the photograph **c**. The initial reference to energy **a** reveals an understanding of the coastal landscape as a system.

**(d)** Changes in sea level resulting from changes in the volume of water in the global oceans are known as eustatic changes. These changes are influenced by changes in global temperatures. This can affect both the amount of water in the oceans and its density. An increase in global temperature leads to higher rates of melting of ice stored on the land in ice sheets, ice caps and valley glaciers. As a consequence there is a global increase in the volume of water in the ocean and a rise in sea level. Also, when temperatures rise, water molecules expand and this also leads to them occupying an increased volume.

ⓔ This is a very clear introduction which clarifies the topic and sets it in a context of climate change. The changes are well explained, with reference to both melting of ice stores and the expansion of warmed water.

> At the end of the last glacial period, which happened about 10,000 years ago, temperatures were about 9°C lower than they are today and sea level was about 90 m lower than it is now. When the sea level rose it formed a number of submerged or flooded landforms.

ⓔ This paragraph gives a clear reference to the past with some data used in evidence, although the figures quoted actually refer to the end of the Würm glacial period, 25,000 years ago.

> One common landform is a ria, which is a drowned river valley. The lowest part of the river's course and the floodplains by the river get completely flooded, but the higher land around it remains dry. In cross section rias have an open V-shape with the valley sides quite gently sloping. In long profile they have a smooth course and water of fairly similar depth. When seen from above they tend to be winding, showing the original route of the river and its valley, formed by fluvial erosion in the river channel and weathering and mass movement processes on the valley sides. There are lots of rias on the south coasts of Devon and Cornwall, for example at Salcombe and Fowey. They are usually about 10 m deep in the middle and provide good, sheltered harbours and are popular for sailing.

ⓔ Rias are well explained and supported with located examples as evidence. The reference to sailing and harbours is not needed.

> Another landform is fjords. Fjords are submerged glacial valleys. They have steep valley sides and the water is very deep. The have a U-shaped cross section because this was the original shape of the glacial valley. They also tend to be much straighter than rias as the glacier would have truncated any interlocking spurs when it moved through the valley. The Sogne Fjord in Norway is 200 km long and over 1000 m deep. The steep mountains around it are 2000 m tall. This was formed after the last glacial period.

ⓔ Again, the explanation is good and the evidence detailed.

> The final landform created by rising sea level is a shingle beach. As sea level rose at the end of the last glacial period, wave action pushed marine sediments onshore, forming tombolos and bars. The tombolo at Chesil Beach was formed during the Flandrian Transgression. Sediment carried into the English Channel by meltwater during the glacial period built up in Lyme Bay. As sea level rose at the end of the glacial period, the sediment was carried onshore by the south-westerly prevailing winds and waves. It moved until becoming attached to the Isle of Portland at one end and the mainland at Abbotsbury at the other. The beach now contains an estimated 100 million tons of shingle, varying in size from pea-sized material to larger pebbles.

ⓔ This is a very well explained final landform with very full use made of the Chesil Beach example.

All of these landforms are also affected by processes going on at present-day sea level and could also change in the future. Both rias and fjords may be altered by the waves acting on their sides and eroding them. The valley sides may also be changed by the subaerial processes in today's climate or in any different climate conditions of the future. This might lead to a reduced steepness of the valley sides.

With sea levels predicted to rise by a further 0.6 m in the next 100 years, water depth in rias and fjords will increase. Marine erosion is also likely to increase due to stormier conditions and larger waves.

Shingle beaches, being made of loose sediment, are especially at risk of modification. The tombolo at Chesil Beach has been changed by present day longshore drift processes and could in the future. Large material has been moved towards Portland by strong winds and waves from southwest.

ⓔ Reference is made to both present-day and future changes to each of the landforms explained earlier. Evidence provided is a little limited, and use could have been made of the data available concerning sediment sizes at either end of Chesil Beach.

In conclusion, it is clear that coastal landforms are influenced by sea-level rises that have occurred in the past which has formed and developed some distinctive landforms such as rias, fjords and shingle beaches. Sea levels are still rising now and will keep rising in the future and so that is influencing them too.

ⓔ **11/16 marks awarded.** A conclusion is provided, which is a summary of the answer.

Overall, there is some high quality explanation of the formation of appropriate landforms and these are supported with appropriate examples and some data as evidence. Unfortunately, there is a lack of consideration of the **relative importance** of the three different time periods concerned. Each is dealt with individually and there is a lack of comparison between them. The student could have made comments about the difference between formation and development in the past, and the modifications happening in the present day. A point could be made about the uncertainty of speculation about the future. Also, it could be noted that it depends how far into the future one is looking.

The student would gain very high marks for AO1 (Knowledge and understanding) = 7/8, but the lack of explicit assessment means that the mark for AO2 (Application of knowledge and understanding) = 4/8. This gives a part (d) total = 11/16.
ⓔ **Total score: 24/33 = a B grade equivalent.**

# Question 2 Glaciated landscapes

**(a)** **Explain how glacier mass balance varies through the year.** [8 marks]

**(b)** **Study Table 2, the number of days per year with freeze–thaw cycles in selected locations.**

**Table 2** The number of days per year with freeze–thaw cycles in selected locations.

| Location | No. of days per year with freeze–thaw cycles |
|---|---|
| Yakutsk, Russia | 42 |
| Fenghuo Shan, Tibet | 354 |
| Tuktoyaktuk, Canada | 43 |
| Mont Blanc Station, Peru | 337 |
| Spitsbergen | 59 |
| Summit Station, Peru | 42 |
| Sonnblick, Alps | 63 |
| Kerguelen Island, South Atlantic | 120 |

**(i)** **Calculate the mean value of the data in Table 2. You must show your working.** [2 marks]

**(ii)** **Calculate the range of data values shown in Table 2. You must show your working.** [2 marks]

**(iii)** **Suggest why the number of days per year with freeze–thaw cycles varies.** [2 marks]

**(c)** **Study Figure 2, a glaciated landscape in the Cairngorms, Scotland.**

**Figure 2** A glaciated landscape in the Cairngorms, Scotland

Explain how glacial erosion has shaped the valley in this landscape. [3 marks]

**(d)* Assess the relative importance of the physical factors influencing glaciated landscape systems.** [16 marks]

Total = 33 marks

(a) This question requires an explanation, but the focus should be on the variation during the year rather than the annual accumulation and ablation totals and their difference.

(b) This is a skills-based question and both mean and range are explicitly mentioned in the specification. Remember to show your working. In (iii) you are not expected to know anything about the specific locations; you just need to suggest appropriate, possible reasons.

(c) There are only 3 marks available here, so keep a clear focus on erosion; no need to mention weathering, for instance.

(d) The command word here is '**assess**' and so you must ultimately decide which of the factors you believe to be the most/least important. The * means that the quality of your extended response will be assessed in this question. You should use full sentences, spell and punctuate correctly and make appropriate use of technical terminology.

### Student answer

**(a)** Glacier mass balance varies through the year due to seasonal differences in temperature and precipitation. In the winter, there is likely to be more accumulation than ablation because temperatures are lower and more precipitation will be in the form of snow, giving large amounts of accumulation. Also the low temperatures mean only small amounts of melting will occur and so ablation will be low. In the summer, temperatures will be higher. Less precipitation will occur and what does fall will be snow rather than rain. There are limited accumulations of ice in the glacier. But as temperatures are higher there will be more melting and so large amounts of ablation will take place.

e **6/8 marks awarded.** This is a competent answer and the student has provided a clear focus on the roles of temperature and precipitation. It is not clear, however, why there is likely to be more precipitation in total during the winter compared to the summer. Also, they haven't referred to the changing balance during the year. In the winter it will be a positive balance with net accumulation, whereas in the summer it will be a negative balance with net ablation.

**(b) (i)** Mean: sum of values = 1060, divided by number of values = 8.1060/8 = 132.5

**e 2/2 marks** The student has correctly calculated the mean value and has shown their working.

> **(ii)** 354 – 42 = 312

**e 2/2 marks**. The student has correctly calculated the range by subtracting the lowest value from the highest, and stating the answer.

> **(iii)** It depends on the variations in temperature between different places. Some locations such as Tibet and Peru have lots of freeze–thaw cycles because of where they are, but others such as Russia and Canada have very few because of their different locations.

**e 0/2 marks**. The student has not really answered the question. They have referred to the data in the table, but have only really offered 'because they are in different locations' as the reason. The key is the different climatic conditions in each location. This influences both the number of diurnal freeze–thaw cycles as well as seasonal variations, with spring and autumn often having more, as temperatures maybe changing from one side of zero to the other.

> **(c)** Moving ice in glaciers has energy and material that enables it to erode the landscape **a**. Glaciers can erode by abrasion, where debris embedded into the bottom and sides of the glacier rubs against the rocks of the valley floor and sides, wearing them away. Glaciers also erode by plucking. This happens when meltwater seeps into joints in the rocks of the valley floor/sides. This then freezes and becomes attached to the glacier. As the glacier advances it pulls pieces of rock away **b**. These processes happen very slowly over long periods of time, especially when the glacier is advancing. Because of its high levels of energy it is able to erode powerfully and the former V-shaped, winding river valley is widened, straightened and deepened to become a U-shaped trough evident in the photo **c**.

**e 3/3 marks**. The student has provided a very full and accurate explanation. There is detailed use of specific process mechanisms **b** and there are clear links to how the valley develops due to erosion. They have also made an explicit link to the photograph **c**. The initial reference to energy **a** reveals an understanding of the glaciated landscape as a system.

> **(d)** Glaciated landscape systems consist of inputs, throughputs and outputs. They are landscapes that have been shaped by the action of glaciers which are large masses of moving ice. There are a number of factors which influence the system, including climate and geology, as well as relief and aspect. Glaciated landscapes form gradually over long periods of time, and there have been a number of glacial periods over the last 2 million years. Some places, such as Britain, show landscape evidence of these earlier glacial periods, whilst others, such as Antarctica, are being actively glaciated now.

ⓔ This is a very clear introduction which clarifies the topic and sets it in a context of a system. The main factors are identified, although latitude and altitude are not mentioned at this stage. A useful temporal and spatial context is also provided.

> Climate is one factor that influences glaciated landscape systems. The amount of snowfall that occurs as precipitation will influence how much ice accumulates in glaciers and the temperature pattern through the year will determine how much ice melts or is lost by sublimation and evaporation. This is the ablation from the glacier. For example, the Antarctic is characterised by sub-zero temperatures throughout the year, typically −50°C in winter to −20°C in summer. This means that there is no melting of snow, no loss of ice to ablation. The ice sheet will continue to grow and increase in thickness due to the excess of accumulation. However, inputs of snow are quite small, with annual totals usually <100 mm per year. This is because Antarctica is under the influence of the Polar Cell, which leads to cold, dense sinking air and so a lack of condensation and cloud formation. In contrast, places such as the Rockies have less extreme temperatures and summer melting occurs as temperatures rise above zero for a few months. Precipitation rates are higher due to the orographic effect and annual totals may be 1500 mm or so. Therefore, this location has high levels of accumulation but also high levels of ablation, so glaciers may grow in the winter, but shrink in the summer, and not really change overall from one year to the next. These differences are because the climate in these locations is itself influenced by their latitude and their altitude.

ⓔ The influence of climate of mass balance is well explained and is supported with contrasting locations, for which evidence is provided. There is a reference to two other factors not mentioned in the introduction. However, there is a missed opportunity here for the influence of climate to be assessed in terms of its importance.

> Another factor is the geology. Glaciers are able to erode the landscape as ice moves forward, especially during periods of growth when they advance. This happens particularly during winter months and when temperatures are getting lower at the start of a glacial period. Glaciers erode by plucking and abrasion but how effective the erosion is does depend upon the resistance of the geology to the erosion processes. Loose, unconsolidated materials such as sands and gravels will be easily eroded whereas hard, well-bonded rocks such as granite will be much more resistant to abrasion. Rocks with lots of joints and cracks may be particularly affected by plucking, as water can enter the joints before freezing and attaching the rocks to the base of the glacier before pulling them out of the ground as the glacier moves forward. Rocks such as limestone have numerous joints and could be affected in this way. The effect of this on the landscape might be that resistant rocks provide the high peaks and arêtes, which are not rapidly eroded, whereas the weaker or jointed rocks provided the deeply eroded troughs. If a valley floor is made up of different geologies, some weaker than others, then the floor may be of very variable depth, with steps between and ribbon lakes occupying the deeper sections, as in the Lake District.

**e** Again, the explanation is good with specific links between geological characteristics and the different erosion processes. Named geologies help provide some evidence, although this is not especially well supported with accurately located examples.

> Other factors such as relief and aspect also have an influence. In areas of steep relief, glaciers will move faster downslope under gravity. This means that erosion rates will be higher and the landscape will become even steeper. Aspect is the direction a slope faces. During the last glacial period in Britain, north- and east-facing slopes received less direct sunlight and so snow tended to accumulate faster and last for longer. This greatly helped the formation of corries by nivation and erosion during rotational movement. In places like the Lake District, there are far more corries on north- and east-facing slopes than there are on south- and west-facing slopes. For example, around Helvellyn there are 3 corries on the north and east side (Nethermost Cove, Red Tarn and Brown Cove) and none on the south and west.

**e** Relief is explained briefly but aspect is very fully and clearly explained, with good supporting evidence from the Lake District. Again there is a missed opportunity here, to explicitly link aspect to microclimate.

> In conclusion, it is clear that glaciated landscape systems are influenced by a number of different factors. These affect how glaciers behave and how they can shape the landscape by their different processes. This can lead to the formation of many different landforms and different types of landscape.

**e** A conclusion is provided, which is a summary of the answer.

Overall, these is some high quality explanation of the role of several factors and the impact of these on processes and ultimately on the landscape system. There are quite a lot of appropriate examples and some data as evidence. Unfortunately, there is a lack of consideration of the **relative importance** of the different factors concerned. Each is dealt with individually and there is a lack of comparison between them. The student could have considered the dominance of climate and the fact that some of the other factors, such as altitude, latitude and aspect, themselves have an influence on the climate, often at a local scale.

The student would gain high marks for AO1 (Knowledge and understanding) = 7/8, but the lack of explicit assessment means that the mark for AO2 (Application of knowledge and understanding) would be less impressive = 4/8.

This gives a part (d) total = 11/16.
**Total score: 24/33 = a B grade equivalent**

## Question 3 Dryland Landscapes

**(a) State how aridity index is calculated and explain why it varies from place to place.** [8 marks]

**(b) Study Table 3, the monthly rainfall for Alice Springs, Australia.**

**Table 3** The monthly rainfall for Alice Springs, Australia.

| Month | Rainfall (mm) |
|---|---|
| January | 42 |
| February | 33 |
| March | 28 |
| April | 10 |
| May | 15 |
| June | 12 |
| July | 8 |
| August | 8 |
| September | 8 |
| October | 17 |
| November | 30 |
| December | 38 |

**(i) Calculate the mean value of the data in Table 3. You must show your working.** [2 marks]

**(ii) Calculate the range of data values shown in Table 3. You must show your working.** [2 marks]

**(iii) Suggest why the monthly rainfall varies.** [2 marks]

**(c) Study Figure 3, a dryland landscape in Death Valley, California.**

**Figure 3** A dryland landscape in Death Valley, California

**Explain how fluvial erosion has shaped the channel in this landscape.** [3 marks]

**(d)* Assess the relative importance of the physical factors influencing dryland landscape systems.** [16 marks]

Total = 33 marks

**(a)** The first part of the question only requires the formula to be stated. The second part of the question requires an explanation, but the focus should be on the variation from place to place rather than over time.

**(b)** This is a skills-based question and both mean and range are explicitly mentioned in the specification. Remember to show your working. In (iii) you are not expected to know anything about the specific location; you just need to suggest appropriate, possible reasons.

**(c)** There are only 3 marks available here, so keep a clear focus on erosion; no need to mention deposition, for instance.

**(d)** The command word here is 'assess' and so you must ultimately decide which of the factors you believe to be the most/least important. The * means that the quality of your extended response will be assessed in this question. You should use full sentences, spell and punctuate correctly and make appropriate use of technical terminology.

## Student answer

**(a)** The aridity index is calculated by working out the ratio of precipitation to potential evapotranspiration. It varies from place to place because it depends on two things, which both vary. The first is precipitation. Some areas have very low precipitation. In places such as the central Sahara Desert it is less than 100 mm per year. This means there is little water provided and the location will be dry. It also depends on the rate of evapotranspiration. This mainly depends on temperature. If temperatures are high, and in the Sahara it gets to over 40°C in the summer, then the small amount of water that is present either gets evaporated, by turning from liquid to gas, or transpired, when it passes through the pores on the surface of the vegetation. So if rainfall is low and temperature is high, then the index of aridity will be very low. Places like this are known as hot arid areas and are very short of water.

**6/8 marks awarded**. This is a competent answer and the student has provided a clear focus on the roles of temperature and precipitation. The processes of evaporation and transpiration are clearly understood and the interpretation of the ratio in the index is correct. However, by only referring to a hot arid location with a low index, they have not addressed the issue of variation successfully. The answer would have benefited from a contrast to a location that was semi-arid.

**(b) (i)** Mean: sum of values = 249, divided by number of values = 12. 249/12 = 20.75 mm.

**e** **2/2 marks awarded.** The student has correctly calculated the mean value and has shown their working.

> **(ii)** 42 – 8 = 34.

**e** **2/2 marks awarded.** The student has correctly calculated the range by subtracting the lowest value from the highest, and stating the answer.

> **(iii)** Rainfall varies during the year. This location is in the southern hemisphere, so winter is June–September. Rainfall is very low in these months, with a minimum of 8 mm. In the summer, October to March, the rainfall is quite a lot higher with a maximum of 42 mm in January. So rainfall varies due to the different seasons.

**e** **0/2 marks awarded.** The student has not answered the question. They have referred to the data in the table but not explained the variations; it is just description. The key is that it depends on the variations in temperature. In summer, e.g. December, the temperatures are higher and so there will be more convectional rainfall than in the winter months, such as August. It may also be influenced by pressure systems. Low pressure systems bring frontal rainfall whilst high pressure systems bring dry conditions. There may be more low pressure systems here in summer.

> **(c)** Running water in rivers has energy and material that enables it to erode the landscape **a**. Rivers can erode by abrasion, where debris carried in the flow rubs against the rocks of the channel bed and banks, wearing them away. They can also erode by hydraulic action, where turbulent flow forces air and water into cracks which widens them and pieces break off. **b**. These processes happen when the river has a lot of discharge, which could be in the wet season and especially after of storm and flash flood. The river than dries up afterwards and so the channel then has no water in it, as at the time in the photo **c**.

**e** **3/3 marks awarded.** The student has provided a very full and accurate explanation. There is detailed use of specific process mechanisms **b** and there are clear links to how the valley develops due to erosion. They have also made an explicit link to the photograph **c**. The initial reference to energy **a** reveals an understanding of the dryland landscape as a system.

> **(d)** Dryland landscape systems consist of inputs, throughputs and outputs. They are landscapes that have been shaped by the action of rivers and wind. There are a number of factors which influence the system, including climate and geology, as well as relief and aspect. Dryland landscapes form gradually over long periods of time, and wind processes act slowly. Some places have landscape evidence of earlier pluvial periods, with higher rainfall, as well as present-day wind action in very arid climates.

**e** This is a very clear introduction which clarifies the topic and sets it in a context of a system. The main factors are identified, although latitude and altitude are not mentioned at this stage. A useful temporal and spatial context is also provided, particularly with reference to pluvial periods.

> Climate is one factor that influences dryland landscape systems. The amount of rainfall that occurs will influence how much water is in river channels and the temperature pattern through the year will determine how much water is lost by transpiration and evaporation. For example, the Sahara is characterised by high temperatures throughout the year, typically 40°C in summer to 20°C in winter. This means that there will be a lot of evapotranspiration and so the rivers will not contain much water. This means they have little energy to carry out geomorphic processes. However, inputs of rainfall are all quite small, with annual totals usually <100mm per year. This is because the Sahara is under the influence of the Hadley Cell, which leads to dense sinking air and so a lack of condensation and cloud formation. However, rainfall does fall, often in heavy, convection storms which can lead to flash flooding. This adds lots of water to rivers and then they will have energy for processes to shape the landscape. Winds are also an important part of the climate factor. Many drylands are very open and flat and so wind speeds can become quite high. Winds are able to carry out geomorphic processes, erosion, deposition and transportation, in similar ways to rivers. Winds can provide inputs of sediment, especially sand, to the system and also take sand away, blowing it into surrounding areas as loess.

**e** The influence of climate on the system is quite clear with references to both energy and material. Temperature, rainfall and wind are all discussed. The influence of climate could be linked to weathering and mass movement processes too. However, there is a missed opportunity here for the influence of climate to be assessed in terms of its importance.

> Another factor is the geology. Rivers and wind are able to erode the landscape as they move. They both do this by abrasion, wearing away the landscape as the material they carry rubs against surface rocks, but how effective the erosion is does depend upon the resistance of the geology to the erosion processes. Loose, unconsolidated materials such as sands and gravels will be easily eroded whereas hard, well-bonded rocks such as granite will be much more resistant to abrasion. Rocks with lots of joints and cracks may be particularly affected by weathering, as water can enter the joints before freezing and thawing. Rocks such as limestone have numerous joints and could be affected in this way. The effect of this on the landscape might be that resistant rocks provide the high mountainous areas, which are not rapidly eroded, whereas the weaker or jointed rocks provided the deeply cut valleys and canyons. If a valley floor is made up of different geologies, some weaker than others, then the canyon sides may be stepped.

**e** Again, the explanation is good with specific links between geological characteristics and the different geomorphic processes. Named geologies help provide some evidence, although this is not especially supported with accurate located examples.

Other factors such as relief and aspect also have an influence. In areas of steep relief, rivers will move faster downslope under gravity. This means that erosion rates will be higher and the landscape will become even steeper. At the bottom of slopes they lose energy and deposition occurs, forming features like alluvial fans and bajadas. These are common in Death Valley, California, where rivers and streams come out of the adjacent mountains and form features 20 km long and 30 m thick. Aspect is the direction a slope faces. Some slopes may face the prevailing wind and be subjected to more erosion. Ventifacts are rocks with one side eroded by wind abrasion into a smooth, flat face. There are a lot of these in the Sahara, typically, several cms in diameter, and sometimes the wind blows them over, exposing another face to the abrasion of the wind which also then becomes smoothed.

**e** Relief and aspect are fully and clearly explained and supporting, located evidence is provided. The use of ventifacts as evidence of aspect is not really relating to slopes, but the principle of exposure to prevailing wind direction is useful. However, the answer again lacks comments about the relative importance of these factors.

In conclusion, it is clear that dryland landscape systems are influenced by a number of different factors. These affect how wind and water can shape the landscape by their different processes. This can lead to the formation of many different landforms and different types of landscape.

**e** A conclusion is provided, which is a summary of the answer. Overall, there is some high quality explanation of the role of several factors and the impact of these on processes and ultimately on the landscape system. There is some use of appropriate examples and some data as evidence. Unfortunately, there is a lack of consideration of the **relative importance** of the different factors concerned. Each is dealt with individually and there is a lack of comparison between them. The student could have considered the dominance of climate and the fact that some of the other factors, such as altitude, latitude and aspect, themselves have an influence on the climate, often at a local scale.

The student would gain high marks for AO1 (Knowledge and understanding) = 7/8, but the lack of explicit assessment means that the mark for AO2 (Application of knowledge and understanding) would be less impressive = 4/8. This gives a part (d) total = 11/16.
**Total score: 24/33 = a B grade equivalent**

# Changing spaces; making places

## Question 1

**(a) Figure 4 shows part of a place in an AC.**

With reference to Figure 4, describe two characteristics which have helped shape the place's profile. [3 marks]

**Figure 4** Part of a place in an Advanced Country (AC)

**(b) Table 4 gives data about two places.**

**Table 4** Data about two places

| | Place A<br>Inner city in an AC | Place B<br>Suburb in an AC |
|---|---|---|
| % long-term unemployed | 5 | 1 |
| % population with no qualifications (e.g. GCSE/A-levels/GNVQ/City & Guilds/degree) | 24 | 13 |
| % of households owner-occupied | 21 | 84 |
| % of households with 1 car or van | 43 | 86 |
| % with long-term health or disability limiting day-to-day activities | 8 | 4 |

**With reference to Figure 4, suggest why social inequalities vary between the two places.** [8 marks]

**(c) Explain how perception influences our sense of place.** [6 marks]

**(d)* With reference to a specific example, assess the relative roles played by different players (stakeholders) in driving economic change in a place.** [16 marks]

e

(a) This sub-part requires a description of two pieces of evidence which must come from the photograph.

(b) This sub-part requires an explanation of the contrasts between the two places. With 8 marks available, you should summarise the contrasts and then explain in detail. Remember to quote data directly from the table.

(c) This sub-part requires explanation of the **link** between perception and sense of place. This is likely to be more convincing if you mention examples to support your points.

(d) The command word here is '**assess**' which means that you must offer some evaluation or judgement. You need to have substantial knowledge of the different players involved and set this clearly in the context of your case study. The * means that the quality of your extended response will be assessed in this question. You should use full sentences, spell and punctuate correctly and make appropriate use of technical terminology.

**Student response**

**(a)** The photograph shows an urban street scene which is densely built up. There is no open space and so its physical characteristic is urban. The local population is more diverse as there is a restaurant which serves curry and this indicates that people from ethnic groups are likely to live in this place.

e **3/3 marks awarded**. The response focuses on the photograph to give two appropriate features which help shape the place profile. It is the mention of building density which is effective in the first characteristic. The second makes direct reference to the resource, linking this evidence with a demographic characteristic.

**(b)** The two places are very different in their populations. Place A is characteristic of many inner city locations while Place B is in the suburbs. Generally Place A is more deprived and suffering from social inequalities than Place B.

e The opening paragraph includes a summary in its final sentence which indicates the student has appreciated the overall contrast between the two places.

About a quarter of the people in Place A do not have any qualifications which makes it much more difficult for them to find jobs. In Place B fewer people have no qualifications, only 13%. This means that long-term unemployment in the suburb is very low compared to the inner city place, 1%:5%. If people are unemployed for a long time, they can often suffer from health issues. Twice as many people in the inner city have long-term health or disability issues than in the suburb 8%:4%. If you are seriously ill or disabled then this can make it very hard to have a job. The wealthier people in the suburbs can afford cars and can buy their own houses. In the inner city more people will be renting as they can't afford to buy.

**e** **6/8 marks awarded.** In the main part of the answer, the second paragraph, several helpful comments are made. The student successfully links together education with employment, quoting figures from the table in support. In addition, a convincing link is made with health matters. The final couple of sentences are relevant but no data from the table is quoted and the comment about wealth is not associated with education or unemployment for example. This is an encouraging response as it keeps its focus on the table of data but requires more detail.

**(c)** Perception plays an important role in how we see places. Perception is subjective as what one person sees is not the same as another person. For example, one person might look at a place such as a centre of a town and be attracted by the clubs and pubs and noise but an older person might not like these. One person will have positive perceptions while the other will have negative ones.

**e** The student clearly understands what is meant by perception and how this can influence a sense of place. The example quoted about the role of age is appropriate.

Gender has an important effect on perception as males and females see the world in different ways. Some places are seen as unsafe by females, such as bus and train stations late at night, although they can also be considered unsafe by some males. It used to be that many public houses were not regarded as where females would be comfortable but this has been changing in the last few years.

**e** Gender is a relevant factor to comment on in the context of perception. The point about some places being perceived as 'safe' or 'unsafe' is convincing.

Religion can be a strong influence on perception. For example, in some cities in India there are separate Muslim and Hindu neighbourhoods. Traditionally, a person from one neighbourhood would avoid going to one with a different religion. The same has been true in some places where Catholics and Protestants have lived in different areas such as Belfast.

ⓔ **5/6 marks awarded**. The paragraph focused on religion needs to comment further about how it influences perception, not just that it does. For example a comment about feeling safe would also be relevant here. Overall this is an effective answer.

**(d)** Economic change involves several different players who all interact. Players are organisations and people involved in a place such as national and local government, businesses and local people.

ⓔ It helps set the scene in these extended prose questions to have an introduction. This student gives an encouraging opening as the term 'player' is clearly understood.

The Birmingham metropolitan region is a large conurbation of just over 1 million people. It became a major manufacturing centre during the industrial revolution making all sorts of metal goods. It became a centre for vehicles with the Austin firm setting up there along with Dunlop tyres. However, in the 1970s it started to decline when industry closed and many people were unemployed.

ⓔ It is good that the student gives some background to their specific example. There is some detail in this paragraph which is focused on economic change, both the initial growth and then more recent decline.

There were several players involved in Birmingham's changes. Overseas car makers such as Nissan and VW began to import more cars into the UK. People were buying these cars and not British ones. There were also TNCs making other metal goods more cheaply than could be done in Birmingham and many small firms went out of business. The inner city of Birmingham also was being redeveloped as much of the housing was poor quality.

ⓔ This paragraph continues to tell the story of economic change including some of the players involved. So far, however, the student has been describing changes rather than assessing the roles of the players. The final sentence is correct but is not related to the question and so adds nothing to the discussion.

In the late twentieth century Birmingham started to be redeveloped. The National Exhibition Centre was built by Birmingham Airport which was expanded. The City Council set up the Heartlands Development Corporation to regenerate some of the inner city areas. The City Council are now trying to attract EU funding from the ERDF and this is important as it is being used to improve housing and build small-scale factories in the inner areas. A large TNC called Tata, who are based in India, now owns Jaguar-Land Rover. They are a very important player as Tata is investing millions of pounds in its factories e.g. at Solihull. This gives jobs to the people working for Jaguar-Land Rover but also for the firms supplying parts for the cars.

ⓔ The answer really begins to 'take-off' in this paragraph. Good factual detail is given showing that the student has substantial knowledge of the example. There is also some assessment with the identification of the importance government and TNCs can play in driving economic change. However, more assessment is required as the answer is still tending to be descriptive.

> Birmingham is also developing other economic activities. The redevelopment of the retail area has made it into one of the largest retail locations outside of central London. This has included the making of Grand Central which is part of the redevelopment of the New Street Station. This has involved several players such as Network Rail, John Lewis department store and many other retailers. This is an important economic change as it employs many people and also improves the look of the area.
>
> There is also a major conference centre and a concert hall. These have been developed by various players and bring jobs to the area.

**e** This section continues to offer good material describing economic change although the comment about the conference centre and concert hall lack details such as which players were involved in their development.

> Transport in Birmingham is receiving investment as the Midland Metro tram system is being developed. This is important as it links the CBD with areas such as West Bromwich and therefore people can use it for commuting. The government are also building HS2 from London to Birmingham. This will cut the journey time to less than 1 hour and possibly bring more investment into the Birmingham region.

**e** Another relevant section regarding economic change.

> There are lots of different players involved in Birmingham's economic change. Most of them bring money for investment and this is important as all development needs this. A very significant player is the City Council and another is TNCs such as Tata. They are all important but some have more power than others.

**e** The discussion concludes with an attempt to summarise, with an interesting final sentence offering assessment. However, in the preceding paragraphs, the student has not made any assessment as to which players have more power. There are also a couple of grammatical errors.

It is encouraging that the student has included some authoritative factual material about economic change in the Birmingham region and has referred to several of the players involved. The knowledge and understanding shown therefore lifts the response into Level 3 in AO1 6/8 marks. The key command phrase in the question, '...assess the relative roles played...' is not explicitly tackled in the discussion. Too often, the student hints at an evaluation of the relative role of a player but does not come out with a clear judgement. In AO2 therefore, the answer is given Level 2 5/8 marks. Overall sub-part (d) receives 11/16.

**Total score: 25/33 = boundary score A/B**

## Knowledge check answers

1 Kinetic, thermal and potential.
2 Friction between the sea floor and the moving water in the wave slows the wave and causes it to steepen. When water depth is < 1.3 x wave height, the wave breaks.
3 It contains numerous small air spaces (pores) in which water can be held.
4 Mass movement involves the force of gravity, whereas transportation requires a moving medium, such as water.
5 Erosion is the wearing away of rock, and so requires a moving force, whereas weathering is the breakdown and decay of rock, and a moving force is not required.
6 Because waves can be refracted around the headland.
7 A process by which salt causes the aggregation of minute clay particles into larger masses that are too heavy to remain suspended in water.
8 It affects the volume of water stored on the land as ice, which in turn affects the volume in the oceans. It also affects the volume in the sea as warmer water is less dense and so occupies a greater volume, and vice versa.
9 Kinetic, thermal and potential.
10 Accumulation – ablation.
11 It contains numerous small air spaces (pores) in which water can be held.
12 Diagenesis: the compression and compaction of snow by the further addition of fresh snow on top.
13 **(a)** basal sliding **(b)** internal deformation.
14 Erosion is the wearing away of rock, and so requires a moving force, whereas weathering is the breakdown and decay of rock, and a moving force is not required.
15 Mass movement involves the force of gravity, whereas transportation requires a moving medium, such as ice.
16 Till is angular, unsorted and stratified whereas outwash is rounded, sorted and stratified.
17 Gradient.
18 A pingo is a domed surface feature, whereas an ognip is a collapsed feature.
19 Kinetic, thermal and potential.
20 The ratio between mean annual precipitation (P) and mean annual potential evapotranspiration (PET).
21 It contains numerous small air spaces (pores) in which water can be held.
22 Exogenic rivers have their source outside of the dryland environment; ephemeral rivers only flow intermittently.
23 Convectional rainfall is common in the summer.
24 Erosion is the wearing away of rock, and so requires a moving force, whereas weathering is the breakdown and decay of rock, and a moving force is not required.
25 Mass movement involves the force of gravity, whereas transportation requires a moving medium, such as ice.
26 There is little vegetation to bind loose sediment together to stop it being carried into rivers.
27 If winds blow from different directions or if the stone is moved to expose a different face to the dominant wind.
28 They are often buried by aeolian or fluvial deposits.
29 Demographic refers to characteristics such as the numbers of inhabitants, what ages they are, numbers of males and females and what is the ethnic composition of the population.
30 Availability of a resource such as a mineral can lead to a place becoming a mining settlement. As long as that mineral is economically valuable the place will prosper but once the mine closes a downward spiral can set in. The location of a university can lead to a place developing as a centre of research and activities such as publishing.
31 A place profile is a description of a place made up of a combination of its physical and human characteristics.
32 Space is the objective meaning of a location such as its map co-ordinates. Place is the subjective meaning of a location such as how you feel about where your home, school or college are. The same space can have different place meanings to different people.
33 A global hub is a place, usually a city, with multiple international connections to other places. Communications, goods, people, ideas and knowledge flow in and out of such a place making it a very dynamic place with an upward economic trend.
34 Informal representations of place are based on subjective opinions. Different people can give very different meanings to the same place based on the same photograph or painting. Some informal representations of place try to emphasise a certain characteristic and so convey a particular image.
35 Relative poverty relates a person's income to local economic conditions. It takes account of the cost of goods and services relative to that in other places. For example, average incomes in one place may seem comparatively high, but if the costs of goods and services are high, then poverty is relative to that situation and can exist at income levels higher than would be the case in places with lower costs of goods and services.
36 Housing often takes a major share of a household's income so differences in quality of housing directly reflect wealth inequalities. Inadequate housing leads to ill-health and this leads to poor educational performance and absence from work and so a downwards spiral is created.
37 Economic change nearly always involves several players, each with their own particular perspective regarding the change. Some may agree in principle, but disagree about the details, such as wanting to build a new road but disagreeing about its actual route. Others may oppose the change completely.
38 Put simply, the 24-hour city never sleeps. Activities such as work, shopping, leisure are happening throughout the day and night. The detail is more complex as some activities tend to dominate during certain periods, e.g. retailing between 9 a.m. and 7 p.m., and offices operate between 8 a.m. and 6 p.m.

Note: page numbers in **bold** indicate defined terms.